Genetic and Environmental Factors during the Growth Period

NATO ASI Series
Advanced Science Institutes Series

A series presenting the results of activities sponsored by the NATO Science Committee, which aims at the dissemination of advanced scientific and technological knowledge, with a view to strengthening links between scientific communities.

The series is published by an international board of publishers in conjunction with the NATO Scientific Affairs Division

A	Life Sciences	Plenum Publishing Corporation
B	Physics	New York and London
C	Mathematical and Physical Sciences	D. Reidel Publishing Company Dordrecht, Boston, and Lancaster
D	Behavioral and Social Sciences	Martinus Nijhoff Publishers
E	Engineering and Materials Sciences	The Hague, Boston, and Lancaster
F	Computer and Systems Sciences	Springer-Verlag
G	Ecological Sciences	Berlin, Heidelberg, New York, and Tokyo

Recent Volumes in this Series

Volume 68—Molecular Models of Photoresponsiveness
edited by G. Montagnoli and B. F. Erlanger

Volume 69—Time-Resolved Fluorescence Spectroscopy in Biochemistry and Biology
edited by R. B. Cundall and R. E. Dale

Volume 70—Genetic and Environmental Factors during the Growth Period
edited by C. Susanne

Volume 71—Physical Methods on Biological Membranes and Their Model Systems
edited by F. Conti, W. E. Blumberg, J. de Gier, and F. Pocchiari

Volume 72—Principles and Methods in Receptor Binding
edited by F. Cattabeni and S. Nicosia

Volume 73—Targets for the Design of Antiviral Agents
edited by E. De Clercq and R. T. Walker

Volume 74—Photoreception and Vision in Invertebrates
edited by M. A. Ali

Series A: Life Sciences

Genetic and Environmental Factors during the Growth Period

Edited by
C. Susanne
Free University of Brussels
Brussels, Belgium

Plenum Press
New York and London
Published in cooperation with NATO Scientific Affairs Division

Proceedings of a NATO Advanced Study Institute on
Genetic and Environmental Factors during the Growth Period,
held August 24-25, 1982,
in Brussels, Belgium

Library of Congress Cataloging in Publication Data

NATO Advanced Study Institute on Genetic and Environmental Factors during
  the Growth Period (1982: Brussels, Belgium)
   Genetic and environmental factors during the growth period.

(NATO ASI series. Series A, Life sciences; v. 70)
  "Proceedings of a NATO Advanced Study Institute on Genetic and Environ-
mental Factors during the Growth Period, held August 24-25, 1982, in Brussels,
Belgium—T.p. verso.
  Includes bibliographical references and index.
  1. Children—Growth—Congresses. 2. Human genetics— Congresses. 3. En-
vironmental health—Congresses. I. Susanne, Charles. II. Title. III. Series. [DNLM:
1. Growth—Congresses. 2. Genetics, Medical—Congresses. 3. Environment—
Congresses. 4. Body height—Congresses. WS 103 N279g 1982]
RJ131.N325   1982   612'.65   83-20193
ISBN 0-306-41479-1

©1984 Plenum Press, New York
A Division of Plenum Publishing Corporation
233 Spring Street, New York, N.Y. 10013

All rights reserved. No part of this book may be reproduced, stored in a retrieval system,
or transmitted in any form or by any means, electronic, mechanical, photocopying,
microfilming, recording, or otherwise, without written permission from the Publisher

Printed in the United States of America

PREFACE

This volume is dedicated to the study of growth and development from the points of view of public health and epidemiology. Many scientists agree that human physical growth and development represent a sensitive response to environmental conditions. From an epidemiological point of view, physical growth and development can be taken as a primary measure of the level of public health or of the quality of the environment.

Reduced growth, smaller body size for instance, can be regarded as a response to adverse environmental conditions, as an indicator of environmental pressures. On the other hand, physical growth and development can also be viewed as development plasticity and as a strategy for human adaptation. However, various environmental effects cannot be isolated from the genetic factors which determine the individual background and the potential limits of growth.

The aim of this volume is to evaluate these genetic and environmental influences as well as their interactions. Methodological work is therefore necessary, including, among others, mathematical methods summarizing the growth curve of an individual and predicting the adult stature from measurements taken during childhood. Of course, methodological aspects are also implied in the way sampling is done and in the way surveys are conducted. And last but not least, methodological aspects are involved in the estimation of heritability during growth and development. It was necessary to review the different tools (univariate and multivariate analysis) capable of estimating the influence of the genetic factors and to illustrate these methodological aspects by results found in anthropometric surveys or in applied fields such as those of growth and physical activity.

Discussing such a broad topic as the impact of environment on human growth and development requires that the boundaries of the review be defined. We focused our interest on two aspects, our understanding of which is less than satisfactory.

Medical aspects are frequently dealt with in studies of growth and development, but the problems related to prematurity and to infants small for gestational age are of major concern not only because of their high incidence but also because of the high cost of adequate care and the poor prognosis when adequate care is absent. The study of intra-uterine growth is thus essential to an understanding and to the prevention of prematurity.

Environmental factors also are too numerous to review completely. Nutrition, infection, high-altitude hypoxia, and cigarette smoking were discussed. But special attention was given to the social environment, the degree of urbanization, the effects of noise on health, and the chronic effects of industrial pollution such as total global air pollution and chronic exposure to lead.

This volume is the result of a workshop held in Brussels in August 1982 with the help of the NATO Scientific Affairs Division. For each of the four chosen topics--methodology, anthropological surveys, medical environment, and ecological environment--one main paper was planned, followed by two or three papers commenting on the review paper or illustrating it by specific examples. A discussion was held at the end of each session and summaries of these discussions are included. The organization of the Workshop proved to be very efficient and is reflected in the structure of this volume.

CONTENTS

METHODOLOGY OF HEIGHT PREDICTION

Predicting the Mature Stature of Preadolescent Children 3
R. Darell Bock

Comments on "Predicting the Mature Stature of Preadolescent Children" 21
Noel Cameron

Discussion Report 29
Ole Lammert

ANTHROPOMETRIC SURVEYS

Methodology of a Cross-sectional Nation-wide Growth Study 37
J.C. van Wieringen and Machteld J. Roede

The Use of Anthropometric Surveys for Studies on the Genetics of Growth 49
William H. Mueller

Methods in Human Growth Genetics 61
Charles Susanne

Effect of Physical Activity on Functional Growth 81
Vassilis Klissouras

Discussion Report 91
Elizabeth S. Watts

MEDICAL ENVIRONMENT

Introduction 99
Machteld J. Roede

Human Growth up to Six Years - Distance and Velocity . . .   101
  Ingeborg Brandt

Medical Environment: Applications in the Field of
    Public Health (Commentary)   123
  Geneviève Massé

Linear and Catch-up Growth in Chronically Ill
    Children   133
  L. Greco and M. Mayer-Greco

Medical Environment - Influences on Growth
    (Commentary)   145
  Olcay Neyzi

Discussion Report   159
  Joan Smith

ECOLOGICAL ENVIRONMENT

Environment and Physical Growth   169
  Anna Ferro-Luzzi

Ecological Environment: Environment and Physical
    Growth (Commentary)   199
  R.J. Rona

Auxological Epidemiology and the Determination of the
    Effects of Noise on Health   209
  Lawrence M. Schell

Effects of Industrial Pollution on Somatic and
    Neuropsychological Development   221
  Roland Hauspie, Marie-Christine Lauwers,
    and Charles Susanne

Discussion Report   235
  Ajit Ray

List of Contributors   243

Index   245

Methodology
of Height Prediction

PREDICTING THE MATURE STATURE OF PREADOLESCENT CHILDREN*

R. Darrell Bock

Departments of Behavioral Sciences and Education
University of Chicago
Chicago, Illinois

Recently, Bock and Thissen (1980) introduced a new statistical method for fitting non-linear growth models to measurements of the height of individual children taken at various times between one year and maturity. This method, called "Maximum A Posteriori" (MAP) estimation, in effect picks from among a parametric family of possible growth curves, that unique curve most probably describing the child's growth, given one or more measurements in this range. The ordinate of this curve at some mature age (e.g., 22 years) then provides a prediction of the stature the child will attain at maturity. Bock and Thissen (1980) show an example of such a prediction, accurate to within 3 cm, based on a single measurement at 9 years of age for an especially tall girl in the Berkeley Guidance Study. They also suggest that this procedure could be improved by incorporating other information such as skeletal age and parental midstature into the prediction.

The present paper explores in greater detail the performance and possible advantages of the Bock-Thissen (BT) method of predicting mature stature. It concentrates primarily on the problem of prediction for girls at age 10 and boys at age 12. The reason for focusing on these ages is that it is in this range that pediatricians most often see children whom their parents consider unusually short or tall for their age. Since it is now possible for pediatricians to intervene medically to alter the mature stature of preadolescent children, there is considerable interest in predictions of stature

---

*I am indebted to Dr. Robert Rosenfield and Dr. Alex Roche for advice concerning this topic.

from information obtained at about these ages. Interventions are undertaken only if the predicted stature is widely deviant relative to prevailing social norms.

1. OTHER METHODS OF PREDICTING MATURE STATURE

The most widely used method of prediction continues to be that of Bayley and Pinneau (1952), based on the child's skeletal age, current height, and chronological age. Skeletal age is typically estimated by comparing hand and wrist radiograms with the developmental atlas of Greulich and Pyle (1959). The Bayley-Pinneau tables give, as a function of bone age and chronological age, the expected percentage of growth remaining for the child. With this percentage and current height, one can solve for the predicted mature stature. Bayley (1962) presents additional tables incorporating parent midstature that improves the predictions somewhat. The improvement is generally more noticeable for younger children.

Other methods based on empirical regression equations have been presented by Tanner et al. (1975) and Roche, Wainer, and Thissen (1975). Tanner's equations were developed from longitudinal data of the Harpenden Growth Study. The Roche-Wainer-Thissen (RWT) procedure is based on similar data from the Fels Institute Growth Study. The accuracy of these three methods has recently been reviewed by Roche (1980). All three methods are roughly comparable in this respect. Standard deviations for predicting mature stature of preadolescent children varies between 2.5 and 4 centimeters depending on sex and the age at which the prediction was made.

The main disadvantage of these procedures is their lack of flexibility. They use the information in a measure of stature at only one time point, and incorporate only the collateral information from bone age and parent midstature. They do not provide any mechanisms for updating their tables or equations to account for secular change in rate and extent of growth, nor take any account of the population to which the child belongs. They do have the advantage of simplicity of calculation; nothing more than a pocket calculator is required. The Bock and Thissen (1980) procedure is more complex, computationally, but it is well within the capacity of a microcomputer. Even pediatricians in private practice could obtain predictions by this method using these inexpensive machines.

2. THE TRIPLE-LOGISTIC MODEL FOR GROWTH AND STATURE

Although the method discussed here could be applied to other parametric models for growth (such as those of Preece and Baines, 1978), the present discussion is limited to the triple-logistic model of Bock and Thissen (1976). That model is based on the idea

METHODOLOGY OF HEIGHT PREDICTION

that growth from infancy to maturity reflects three cycles of hormonal influence. The main cycle is thought to be controlled by thyroid and growth hormones; it begins between one and two years of age, after the influence of fetal hormones has subsided, and continues in some degree until maturity. Superimposed on this main cycle is a minor mid-childhood cycle possibly due to the maturing of the adrenal cortex (adrenarchy). Finally, there is the major adolescent cycle controlled by the maturing of the gonads and the increased levels of testosterone in males and estrogen in females. (See Rosenfield, 1979.)

In the triple-logistic model, each of these cycles is represented by a logistic growth function with its particular location, slope and size parameters. Since these parameters are individual characteristics, they must be estimated from the growth measurements on each child. Methods for their estimation incorporating other information about the child such as parental mid-stature, skeletal age, secondary sex characteristics, and population membership are discussed below.

In the parameterization of Bock and Thissen (1976), the triple-logistic model is as follows:

$$y = a \left[\frac{1 - p}{1 + \exp[-b_1(x - c_1)]} + \frac{p}{1 + \exp[-b_2(x - c_2)]} \right] + \frac{a_3}{1 + \exp[-b_3(x - c_3)]}, \quad (1)$$

where

- a = the contribution of the main cycle to mature stature
- b_1 = the slope of the initial component at its maximum velocity
- c_1 = age of maximum velocity of the initial component
- b_2 = the slope of the midchildhood component at maximum velocity
- c_2 = age at maximum velocity of the midchildhood component
- p = proportion of the main cycle attributiable to the mid-childhood component
- a_3 = the contribution of the adolescent component to mature stature
- b_3 = the slope of the adolescent component at maxiumu velocity
- c_3 = age at maximum velocity of the adolescent component

In the process of fitting the parameters of this model to data from the Berkeley Growth Study, Bock and Thissen found that the nine parameters were not independent in that data. It appeared that, both in boys and in girls, the parameter p could be expressed as a linear function of the location and slope of the infantile and mid-childhood logistic components as follows:

$$p = 0.851 + 0.041b_1 - 0.018c_1 - 0.517b_2 - 0.042c_2 \qquad (2)$$

On the basis of (2), the parameter p can be excluded from (1) and the remaining eight parameters estimated.

Examples of the triple-logistic model (1) fitted to data under restriction (2) by the method of Bock and Thissen (1980) are shown in Figure 1. The curves in Figure 1 describe the growth of a boy and girl of average stature for their sex from the Berkeley Guidance Study (cases 220 and 383 in Tuddenham and Snyder, 1954). The measurements at yearly intervals between 2 and 8 and half-year intervals elsewhere are shown as solid circles about two standard errors in diameter (assuming a measurement error of 0.5 cm). The solid line is the growth curve between one and nineteen years of age fitted to these points.

The empirical growth velocities (change in measured stature divided by change in age) by years between 2 and 8 and half years elsewhere are shown as open circles. (They are plotted relative to the velocity scale at the right.) The broken line represents the point velocities obtained by differentiating the fitted triple logistic models with respect to age.

Both the growth curves and the velocity curves fit the data well over this range. As tends to be generally the case in both the Berkeley and Fels data, the goodness-of-fit for the girl (lower panel) is better than that for the boy (upper panel). It is hard to imagine how fit for the girl, either for stature or velocity, could be improved.

The boy, one the other hand, shows more instability of growth, especially in the early years, seen most clearly in the empirical velocities, which fluctuate about the trend line more markedly than those of the girl. The fitted curve for the boy also appears to decelerate somewhat more rapidly than observed growth in late adolescence. At the important periods of the preadolescent minimum and adolescent maximum growth, however, the fits are excellent for both the boy and the girl. The triple-logistic model provides a convenient and accurate graduation of the growth measurements for purposes of estimating these and other descriptive features ("biological" parameters) such as the midchildhood maximum, and for estimating smoothed velocities. In this connection, note that children represented in Figure 1 show little if any evidence of the midchildhood growth spurt. In the Berkeley and Fels data, about half of the cases exhibit a detectable spurt with the maximum occurring at about age 6 1/2 years in boys and 5 1/2 years in girls. The growth of the boy and girl in Figure 1 is typical of those cases that do not show the midchildhood spurt.

As we shall see elsewhere in this paper, the fit of the triple-logistic model is equally good in more deviant growth patterns. In particular, the cases discussed in Section 4 and illustrated in Figures 4-6 while showing retarded or precocious adolescent growth that complicates the prediction of mature stature, do not fit the triple-logistic model any less well. No cases that could be considered failures of fit, when the Bock and Thissen (1980) method was used, have been found in the Tuddenham and Snyder (1954) data or in the Fels cases so far analyzed. The model is, of course, capable of representing only strictly increasing height during the period of growth. It could not represent growth arrest in failures to thrive, but it might very well describe the pattern of growth before and after arrest.

3. COMPARISON OF THE BT AND RWT METHODS OF PREDICTION

As a preliminary assessment of the efficacy of the BT method, the deviations of predicted from actual mature stature have been compared for the BT and RWT methods on the same subjects from the Fels Institute Growth Study that were used in the development of the RWT method. (We are indebted to Howard Wainer for making these data available.)

For present purposes, predictions by the BT method used either height at or near the 10th and 12th birthday in girls and boys, respectively, or height plus skeletal age at these chronological ages. The RWT predictions use height and weight at 10 or 12, skeletal age and parental midstature.

Inasmuch as the RWT method is being evaluated on the same data with which it was developed and also incorporates parental midstature, these comparisons may not seem entirely fair to the BT method, which was developed with the Berkeley data and does not in the present application employ weight or parental midstature. Roche, Wainer and Thissen have shown, however, that their method works equally well in the Fels, Berkeley and Denver data, suggesting that it is not "capitalizing on chance" in the Fels data, and they also find that parental midstature contributes very little to prediction after midchildhood. Weight is also less important as a predictor and is not included in systems other than RWT. Thus, comparisons of the two methods in these data seems reasonable.

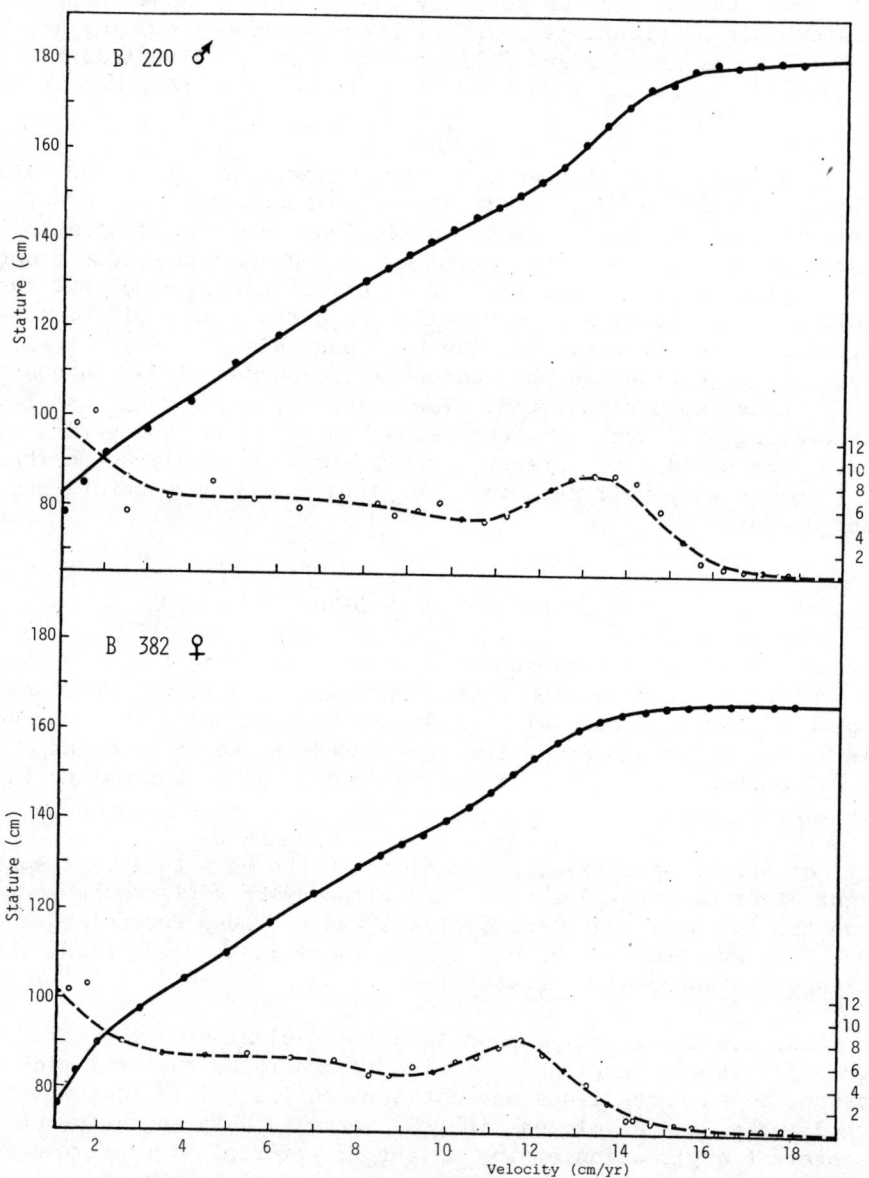

Fig. 1. Observed heights (closed circles) and velocities (open circles) for two children with average mature stature taken from the Berkeley Guidance Study. Fitted triple-logistic growth models (solid lines) and derived velocities (broken lines).

METHODOLOGY OF HEIGHT PREDICTION

From the pediatricians point of view, the success of predictions of mature stature might be judged by the relative frequency with which they fall within plus or minus 5 cm of the true value. Discrepancies greater than these limits would undoubtedly be cause for concern to both the pediatrician and to the child's parents, while smaller errors would rarely be distinguished among inside this range.

On this criterion, the RWT method performs better for the Fels boys than does the BT method. As shown on the top panel of Figure 2, the BT predictions without skeletal age err by more than plus or minus 5 cm in 22 out of 98 cases. Inclusion of skeletal age (by the method discussed below) does not improve matters; the middle panel of Figure 2 shows 23 errors in excess of 5 cm when skeletal age is included. The RWT method (bottom panel of Figure 2), on the other hand, shows only 7 discrepant cases out of 98.

For the Fels girls, however, the situation is quite different. As shown in Figure 3, inclusion of skeletal age (middle panel) reduces the number of discrepant cases by the BT method from 14 out of 83 to 11. It then performs as well as the RWT method (bottom panel), which also shows 11 discrepant cases.

The accuracy of the BT method depends upon the suitability of the description of the population distribution of the growth parameters in the triple logistic model. In the present calculations, this distribution was assumed to be multivariate normal with mean vectors and covariance matrices estimated provisionally from the Berkeley boys' and girls' samples. Somewhat better predictions would be obtained if the means, variances and covariances were calculated conditional on the parental midstature and possibly the current weight for the child in question. This would specialize the predictions to the particular subpopulation from which the child is drawn and would exploit any linear relationship that exists between current weight and mature stature.

The BT method is also influenced by the assumed measurement-error variance. The results shown in Figures 2 and 3 assume a measurement-error standard deviation of 0.5 cm. This figure appears to be roughly in accord with variations that are seen in the annual measurements of the Fels subjects after they have reached maturity. Alternatively, the measurement error variance could be chosen to <u>maximize</u> the accuracy of predicting mature stature from measurements taken in preadolescence.

If these steps improve the predictions of mature stature in boys, the BT method would compare favorably with any of the presently existing methods of height prediction. They would now, however solve the problem of the failure of prediction that afflict all the methods. In these Fels subjects, for example, the cases for which

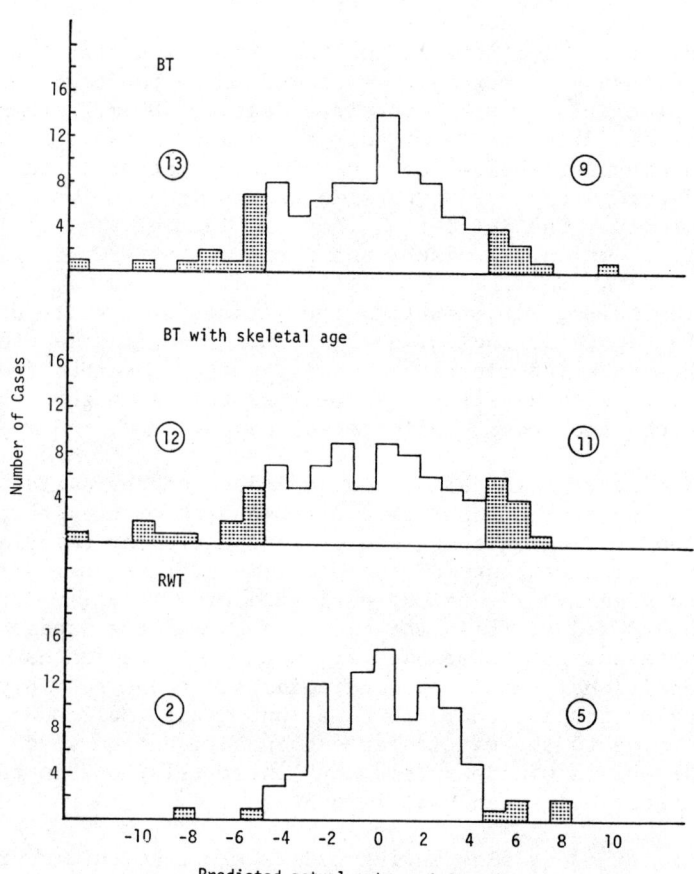

Fig. 2. Errors of predicting mature statures of Fels boys (N=98) from one measurement of height at age 12 years. Top: Bock-Thissen method. Middle: Bock-Thissen method substituting skeletal age for chronological age. Bottom: Roche-Wainer-Thissen method.

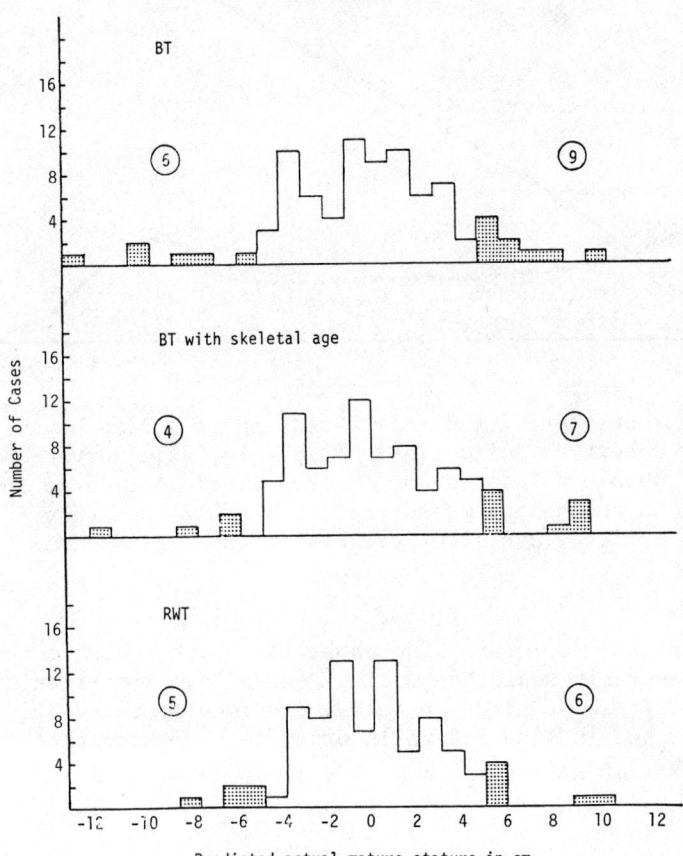

Fig. 3. Errors of predicting mature statures of Fels girls (N=83) from one measurement of height at age 10 years. Top: Bock-Thissen method. Middle: Bock-Thissen method substituting skeletal age for chronological age. Bottom: Roche-Wainer-Thissen method.

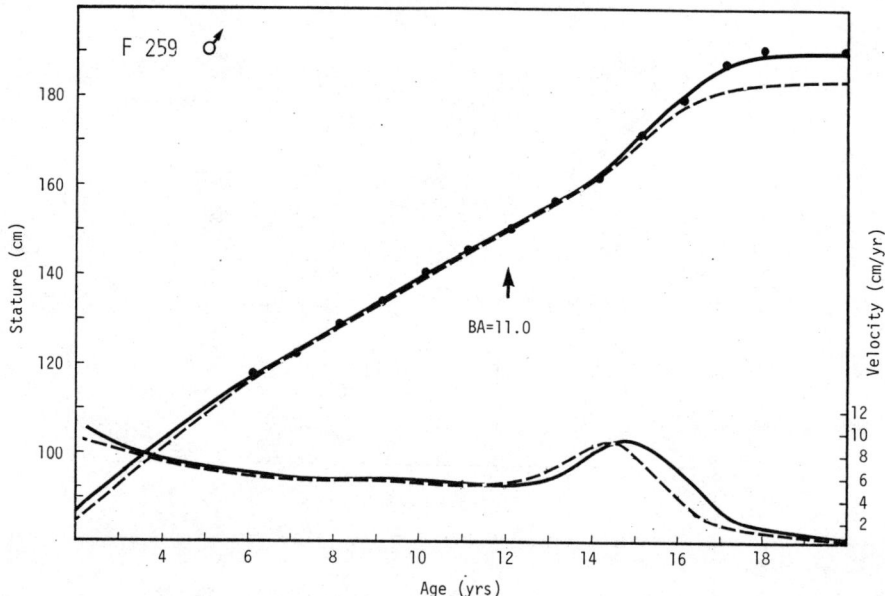

Fig. 4. Poorly predicted growth pattern of a Fels boy with delayed puberty. Solid line: Triple-logistic curve conditional on all points shown. Broken line: T-L curve conditional on one height measurement at age 12, but substituting skeletal age 11.0 years for chronological age.

the RWT method fails are also those for which the BT method fails, as so presumably would the Bailey-Penneau and Tanner-Whitehouse methods. It is of interest to examine some of these difficult-to-predict cases in more detail in order to see where the BT methods go astray.

4. PATTERNS OF HARD-TO-PREDICT GROWTH

Figure 4 shows MAP fitted triple-logistic growth and velocity curves for a tall boy (Fels case No. 259) with an atypically late and protracted adolescent growth spurt. The solid lines in Figure 4 are the curves fitted to the annual measurements shown as small black circles, about two standard errors in diameter. The broken lines represent triple-logistic growth, and derived velocity, fitted conditional on the single measurement of height (150.8 cm) at age 12. The skeletal age of this boy at CA 12 was 11 years. In fitting the curve, therefore, the skeletal age was substituted for chronological age. This gives the curve that would be expected if the boy had attained the height of 156.1 cm at 11 years. This, in effect, gives the expected curve an additional year of growth and brings it in better conformity with the actual growth. With skeletal

METHODOLOGY OF HEIGHT PREDICTION   13

age accounted for in this way, however, the predicted growth of this boy in the range 17 to 20 years still falls short of the actual growth.

Another pattern difficult to predict at these ages is a short, sharp spurt occurring soon after adolescence. Examples are shown in Figures 5 (top) and 6. A possible explanation for these various types of hard-to-predict growth curves may lie in genetically controlled growth patterns. Bock and Thissen (1975) have discussed two sets of triplets from the Fels study in which the two identical twins showed remarkable similar patterns of growth at all ages, whereas the fraternal twin showed a distinguishably different pattern.

In search of evidence of familial effects in hard-to-predict growth patterns in the Fels sample, which contains many siblings, we may focus on the families of a boy and girl, both of whom were short and grossly underpredicted on the basis of their respective height measurement at 12 and 10 years. The fitted and predicted growth and velocity curves for the index case (Fels No. 378) and his younger brother (No. 464) are shown in the top and bottom panels of Figure 5. The index case was 145.2 cm at age 12, at which time his skeletal age was also 12 years. In spite of this skeletal age reading, however, it is apparent from the fitted growth curve that this boy had already nearly reached his adolescent velocity maximum, and his growth in stature and in skeletal age quickly matured and were nearly at their final level by age 14--very early for a boy. Unfortunately, this combination of height and skeletal age at chronological age 12 is far more typical of the growth pattern of a much taller boy, for whom maximum adolescent velocity is expected at 14.5 years. Having nothing more to go on, the prediction of mature stature by any method would greatly overestimate the actual value of this case.

The brother (Fels No. 464) of this boy, on the other hand, has a growth curve close to that expected of a boy who attains a height of 142.5 cm at 12 years. His mature stature is therefore predicted almost exactly. There is no evidence in these brothers of a genetic adolescent precocity in this family.

The growth and velocity curves of his sisters depicted in Figure 6 show a different picture, however. Both the index case (Fels No. 37) and her younger sister (Fels No. 127) are clear cases of precocious adolescence, and their mature statures are grossly overpredicted. Their bone ages at chronological age 10 are advanced by six months and one year, but this is not enough to improve prediction markedly. Since it is difficult to imagine in these middle-class Ohio families any environmental influence strong enough to produce these deviant growth patterns, we must suppose that the two sisters share most of the genes controlling the timing of adolescent development, whereas the two brothers, who could by chance

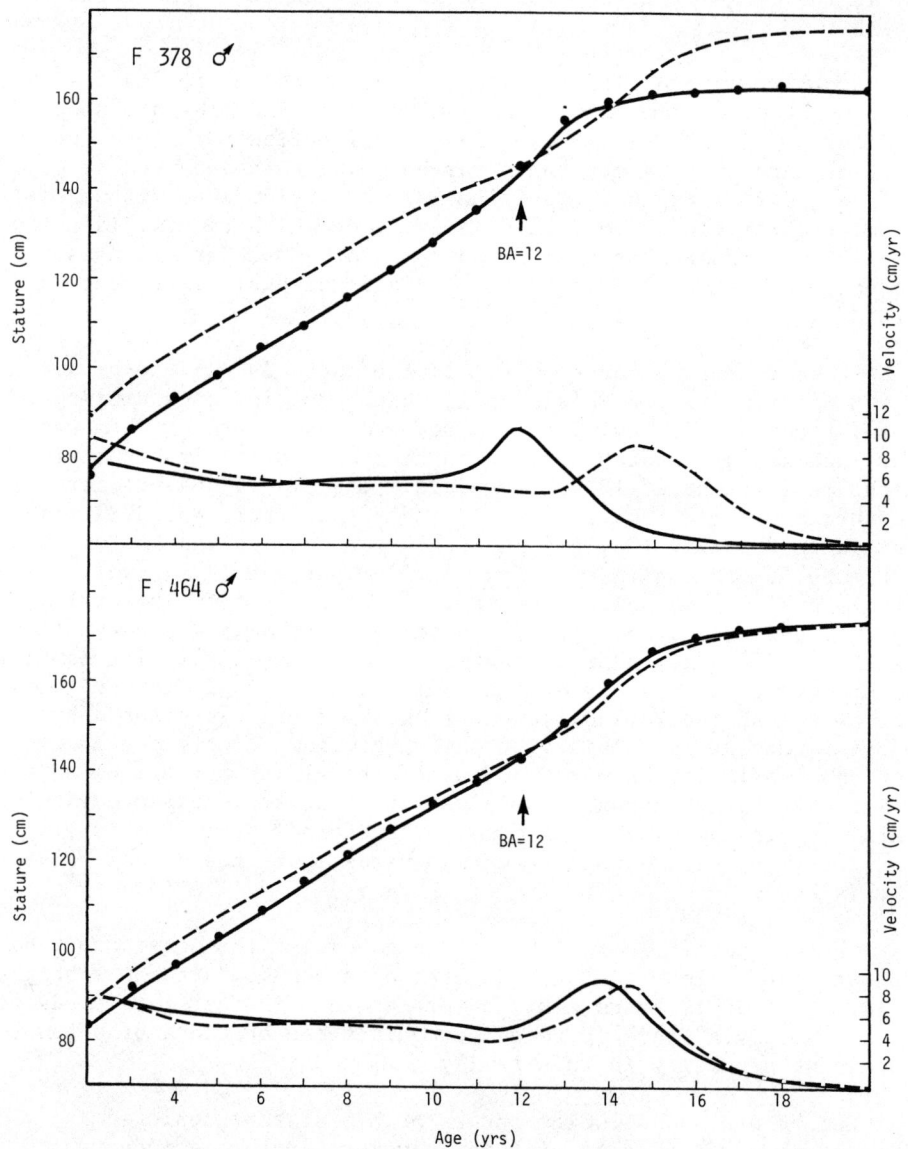

Fig. 5. Poorly predicted Fels boy (top panel) with precocious adolescent growth but normal skeletal age at 12 years. Normal, well-predicted brother of the above (bottom panel). Solid lines: T-L models conditional on the points shown. Broken lines: T-L models conditional on one height measurement at age 12.

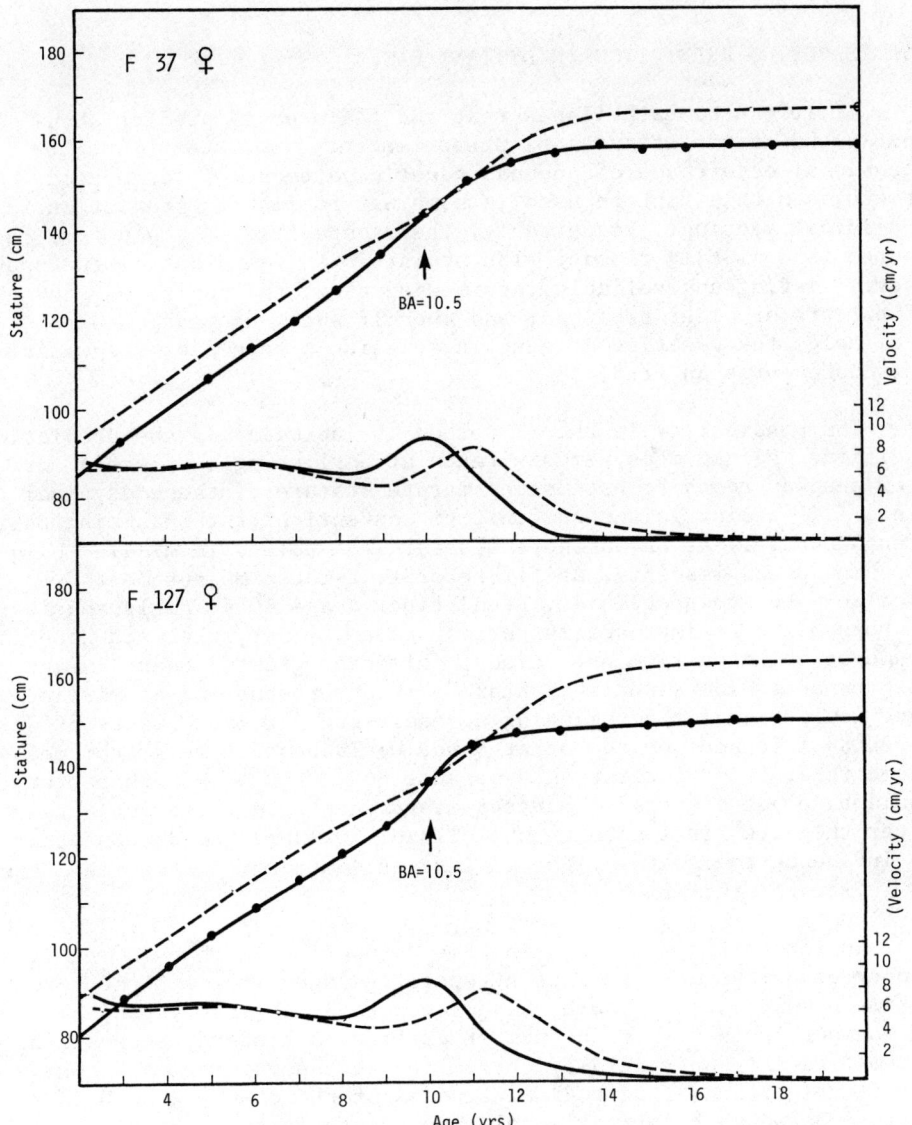

Fig. 6. Poorly predicted Fels girl (top panel) with precocious adolescent growth on one-half year acceleration of skeletal age at 10 years CA. Poorly predicted sister of the above (bottom panel). Solid lines: T-L models conditional on all points shown. Broken lines: T-L models conditional on one height measurement at 10 years and with skeletal age substituted for chronological age.

have relatively few genes in common, were discordant in this trait.

5. IMPROVING PREDICTION IN DEVIANT CASES

If it can be established that the failures to predict accurately the mature statures of preadolescents is traceable to the occasional occurrence of unusual genetic patterns of adolescent growth, can this fact be used in some way to improve prediction in a clinical setting? Regrettably, the prospect of any help from this source is decidedly cloudy. The pediatrician could not depend upon there existing any reliable information about the course of growth of parents or older siblings, and even if such information were available, the vagaries of gene segregation within sibships would introduce large uncertainties.

The possibility in the BT method of including in the prediction additional height measurements taken at earlier ages offers a more promising approach to estimating mature stature of these atypical cases. If pediatricians had uniform conventions for measuring children accurately at one or more standard age points in middle childhood and preadolescence, and if records of these measurements were available to growth clinics, prediction at age 10 in girls and 12 in boys might be improved in certain cases. Most likely to benefit would be children with precocious adolescent growth, where larger than expected increases in stature between present and at earlier ages would signal the beginning of adolescent growth. Cases of delayed adolescence, on the other hand, would benefit less from measurements at these earlier ages because no change of growth pattern would have yet occurred. Further measurements on these subjects after they are first seen in the clinic, however, would soon begin to influence the shape of the adolescent component in the model and thus improve prediction.

An interesting question in this connection is what is the least number of height measurements at specified ages are sufficient to estimate an accurate growth curve for normal subjects. Probably the number is less than the number of free parameters, because the parameters are substantially correlated in the population. Conservatively, then, longitudinal studies of children's growth in stature based on 8 optimally selected points in the age range 1 to 18 years should serve almost as well as annual measurements to describe individual growth in the population of interest. For the description of growth in special populations or subpopulations, longitudinal studies considerably less costly than those conventional studies based on annual and semiannual measurements would then be possible.

Another approach to better prediction is through improved techniques of measuring growth and development. In each of the cases of deviant adolescent growth shown in Figures 5 and 6, the growth velocities, calculated from the triple-logistic curve fitted to all the data, shows that, at age 12 in cases 251 and 378 and at age 10 in cases 37 and 127, the growth velocities are more accurately expressing the actual developmental age than is the skeletal age. It may be, of course, that the skeletal age is inaccurate and that improvements in its determination would improve prediction. There will soon be an opportunity to test this hypothesis: Dr. Alex Roche and David Thissen are presently engaged in applying modern item-response estimation of developmental age from objective ratings of multiple ossification sites in hand-wrist radiograms of the Fels subjects. Their refined estimates of skeletal age may improve prediction in the deviant Fels cases.

The other possibility is to increase the accuracy estimates of growth velocity in relatively short time intervals. Since it is only necessary to measure change within each subject, absolute measurement of overall stature is not necessary. To detect the adolescent increase in velocity, for example, it would be advantageous to measure trunk growth, which accounts for more of the adolescent increase than does limb growth.

One could imagine a technique such as follows. When the child is first seen, a 0.5 mm dot of a colorless fluorescent compound is tattooed at the base of the sacrum and at a vertebra between the shoulder blades. The child is then positioned supine upon a table with a plexiglass panel down the middle. The child's knees are raised and head supported so that the back is in smooth contact with the transparent panel. An optical device with an ultraviolet light source is then moved along a metric track, centered successively on the two dots, and the distance read. The procedure is repeated at the same time of day after 3, 6 and 9 weeks. Simple linear least-squares is then used to estimate from the four points the rate of change of trunk length with age. It should be possible by this method to detect growth of as little as 4 mm in 9 weeks or about 2.3 cm per year.

With this velocity estimate, plus the measures of overall standing height or recumbent length taken at the same times (and possibly a determination of skeletal age), the triple-logistic model could be fitted conditional on height and tangent to the estimated linear velocity curve at the given chronological ages. Thus, the increase in growth velocity at the beginning of adolescence, which leads the age of maximum velocity by more than a year in even the most precocious adolescents, could be detected at the earliest possible age. This information would clearly distinguish cases such as Fels No. 37, who at 10 years C.A. is near adolescent peak height velocity from his brother, No. 27, who at the same age is still at the preadolescent level.

These discriminations could, of course, also be aided by ratings of other sexual characteristics such as pubic hair, breast development, testes length, etc., or even direct assays of the various hormones influencing growth. At present, however, the measurement error and temporal instability of these indices are so much larger than those of gross anatomical measurement that they do not add much to accurate prediction. Barring some breakthrough in pediatric endocrinology, refinements of anthropometric techniques may be the most promising approach to detecting and overcoming the hard-to-predict patterns of adolescent growth in stature.

CONCLUSIONS

The triple-logistic model provides a convenient and highly accurate representation of normal growth in stature and growth velocity between one year and maturity. Using the maximum a posteriori (MAP) method, the model can be fitted dependably to any number of observed measurements. When fitted by this method to a single measurement at age 10 in girls and age 12 in boys, the model provides, when bone age is taken into account, prediction of the mature stature of girls that is equal in accuracy to the Roche-Wainer-Thissen (RWT) regression method. In boys, this method is not as accurate as RWT, but may become so when other information such as weight and midparent height are taken into account.

Prediction by any of the currently used methods fails badly for certain atypical growth patterns, however. Some way of identifying these cases, possibly through better assessments of development age or more accurate measurement of growth velocity, are needed to attain the virtually one hundred percent confidence in growth predictions that is required in pediatric practice.

REFERENCES

Bayley, N., 1962, The accurate prediction of growth and adult height, Modern Problems in Paediatrics, 7:234.
Bayley, N., and Pinneau, S. R., 1952, Tables for predicting adult height from skeletal age: Revised for use with Greulich-Pyle hand standards, J. Pediatrics, 40: 423-441.
Bock, R. D., and Thissen, D, 1980, Statistical problems of fitting individual growth curves, in: "Human Physical Growth and Maturation: Methodologies and Factors," F. E. Johnston, A. F. Roche and C. Susanne, eds., Plenum, New York.
Greulich, W. W., and Pyle, S. I., 1959, "Radiographic Atlas of Skeletal Development of the Hand and Wrist, 2nd Ed.," Stanford University Press, Stanford.

Preece, M. A., and Baines, M. J., 1978, A new family of mathematical models describing the human growth curve. Ann. of Human Biol., 5:1-24.
Roche, A. F., 1980, Prediction, in: "Human Physical Growth and Maturation: Methodologies and Factors," F. E. Johnston, A. F. Roche and C. Susanne, eds., Plenum, New York.
Roche, A. F., Wainer, H., and Thissen, D., 1975, The RWT method for the prediction of adult stature, Pediatrics, 56:957.
Rosenfield, R. L., 1979, Somatic growth and maturation, in: "Endocrinology, Vol. III," Grune and Stratton, New York.
Tanner, J. M., Whitehouse, R. H., Marshall, W. A., Healy, M. J. R., and Goldstein, H., 1975, "Assessment of Skeletal Maturity and Prediction of Adult Height," Academic Press, New York.
Tuddenham, R. D., and Snyder, M. M., 1954, Physical growth of California boys and girls from birth to eighteen years, Univ. of Calif. Publications in Child Develop., 1:183-364.

COMMENTS ON "PREDICTING THE MATURE STATURE OF PREADOLESCENT

CHILDREN" BY DARRELL BOCK

    Noel Cameron

    Department of Growth & Development
    Institute of Child Health
    University of London

Bock presents an approach to adult height prediction based on fitting the triple-logistic growth function (Bock & Thissen, 1980) to individual height data. It is of importance, both to understanding the implications of Bock's approach and to the following comments, that the significance of the Maximum A Posteriori (MAP) estimation of function parameters is realised. The present use of growth functions or models is made possible only in the presence of suitable data. These data must have more individual data points than the number of parameters in the function to be fitted. For a Preece-Baines Model 1 fit for example (Preece & Baines, 1978) at least 6 data points must be present to fit the 5 parameter model. Similarly for the 9 parameter triple-logistic (Bock & Thissen, 1980) at least 10 data points must be present. The satisfaction of these criteria poses obvious problems to research workers wanting to predict adult stature, interpolate or extrapolate data or derive biological parameters, such as peak height velocity, from functions fitted to few data points

The MAP estimation makes these problems far less important because theoretically only one data point is necessary to fit any individual child as long as some information is already known about the specific population from which the child originates. Obviously with more data points the goodness of the fit improves but it is no longer necessary to have a greater number of data points than model parameters.

Although height prediction is the main point of Bock's paper his techniques have far greater implications in other areas of analysis and description. I am grateful, therefore, that my brief in this symposium is to comment on Bock's technique in terms of its function, its relationship to longitudinal growth studies and finally its relationship to other growth models.

FUNCTION

It is apparent that, within the context of this paper, the function of Bock's approach is to obtain accurate estimates of mature stature from one measurement occasion. The viewpoint is expressed that paediatricians judge the success of a prediction technique by the frequency with which the predictions fall within ±5 cm of the actual value. Thus the comparisons made between the Bock-Thissen (BT) technique and the Roche-Wainer-Thissen (RWT) technique (Roche, Wainer & Thissen, 1975) are based on the number of children falling outside these limits. There is no reason to suppose that the limits of ±5 cm have any great importance from the paediatricians point of view. Indeed these limits are those they have been told to expect from the developers of the different prediction techniques. As far as the paediatrician is concerned the success of a prediction technique must be how closely the predictions vary about actual height. They would, I am sure, be far happier with limits of less than ±5 cm and to the patients the accuracy with which we predict is of far more personal interest. Whilst ±5 cm are reasonable statistical limits to judge accuracy we must not neglect the pattern of accuracy within these limits. To investigate whether the BT and RWT techniques "compare favourably" therefore I have subdivided Bock's result into limits of ± 2, 3, 4 and 5 cm. In this way we obtain a more detailed description of the overall success of the technique.

Table 1 demonstrates that for boys the RWT technique is at least 20% better than the BT technique at ± 3, 4 and 5 cm and 11% better at 2 cm. The sample size is not constant at 98 cases but for the situation in which skeletal age is used for the BT technique it is in fact 88 cases and reverts to 98 cases for the RWT technique. Thus percentages rather than actual numbers must be used for the comparison.

The situation is "quite different" as Bock suggests, for the girls but it is only "favourable" at the ±5 cm limits. Within this the percentage of subjects within ±2, 3 and 4 cm are 5%, 12% and 9% better respectively of the total sample for the RWT technique. 16% of the sample are predicting within the ±4-5 cm range, i.e., worse than 4 cm better than 5 cm for the BT technique.

I would take the view, therefore, that RWT is still a significantly better prediction technique and have yet to be convinced that the complex mathematics required to use the MAP with the triple-logistic are not a major "stumbling block" for the acceptance of a prediction technique that is not significantly better than existing methods.

In deference to Bock it is not necessarily the mature stature that is the most important outcome of this work. Indeed the

TABLE 1

BOYS	R-W-T N=98		B-T + Skel Age N=88	
Limits	n	% of total	n	% of total
±2 cm	45	46	31	35
±3 cm	69	70	44	50
±4 cm	83	85	54	61
±5 cm	91	93	65	74

GIRLS	N=83		N=83	
Limits	n	% of total	n	% of total
±2 cm	38	46	34	41
±3 cm	54	65	44	53
±4 cm	68	82	61	73
±5 cm	72	87	71	86

R-W-T and B-T prediction techniques compared at different limits of accuracy.

elucidation of adolescent parameters such as take-off, peak velocity etc. might be far more interesting particularly when data is missing. Also of course the Bock approach gains considerably when more than one data point is available whereas the other major techniques can only take account of the single data point and at present use no information regarding actual or inferred velocity. If for example we predict the height of a girl at age 10, 11, 12 all we can do is say the three predictions agree well (or badly) and take some sort of average. But Bock could incorporate the 10 and 11 year data and arrive at a 12 year prediction which correctly sums it all up.

LONGITUDINAL GROWTH STUDIES

The main problems associated with analysing longitudinal growth studies concern missing data, early finishing and late starting subjects, non-standard age intervals or as Goldstein (1979) calls them "variable occasion times" and, at times, the lack of maturity or somatic anchoring points that may be used to "centre" the data.

Missing data within a series of examinations of an individual forces the researcher to use a mixed longitudinal analysis design such as the Patterson technique (1950). Whilst this technique exists in relatively manageable computer programs the necessity of its use detracts from the longitudinal objectives of a study. Various techniques have emerged in recent years to fit short-range curves to longitudinal data to obtain a reasonable missing value rather than the usual strategum of linear interpolation. Karlberg and his colleagues in Sweden (Karlberg et al, 1976) for instance, fitted a second-degree polynomial to the data points either side of the missing value to obtain an estimate of the new value. In London we have routinely fitted quadratic equations to obtain missing values and latterly have used the Preece-Baines function to fit the full extent of the individual's data and thus obtain a missing value. For the latter technique to work it is necessary to have a number of critical data points - essentially those around take-off and peak-velocity and preferably some in childhood.

The advantage of Bock's technique is that information from all subjects is used to improve the estimates for each individual.

Early finishers and late starters in longitudinal studies reduce the accuracy of the end age ranges in longitudinally based standards. The lack of uniform school attendance age in some countries and a varied school finishing age in most countries, means that school-based samples are subject to untidy early age ranges and high attrition rates at late ages. An obvious advantage of Bock's technique is to improve the amount of available data at these age ranges. The Bayesian approach, as we have seen allows the incorporation of other variables to improve the final fit and therefore the accuracy of extrapolated values.

Non-standard age intervals are present in almost every longitudinal study because of the impossibility of seeing a child exactly at its birthday or mid-year day. The fitting of any function allows data at exact age points to be obtained. It must be remembered that the time between measurement occasions remains constant so that seasonal or other biological effects do not lead to different growth patterns for different individuals (Goldstein, 1979).

The lack of supporting maturational or developmental data poses serious problems when attempting to overcome the "phase effect" (Boas, 1892) and apply a <u>longitudinal</u> analysis to longitudinal data. The use of peak height velocity as a "centering" point for individual velocity curves has been demonstrated many times (e.g., Tanner et al, 1966) and has become an accepted technique to view the relationship of other somatic variables to general adolescent growth (Tanner et al, 1981). Preece and Baines (1978) used their function to gain information on various "biological parameters" relating to adolescence such as "take-off" and "peak-velocity" ages and, in my view, it is the availability of this type of information that makes fitted functions an attractive analytical or elaborative procedure. Once again the Bayesian approach allows the estimation of such biological parameters from very few data points and not necessarily any within the adolescent period. Thus it might be possible, for instance, to make much greater use of studies that have finished prematurely without an adolescent component. I think particularly of a large longitudinal study in Quebec that studied the growth and physical activity of children between the ages of 6 and 12 years (Lavalle, 1981, unpublished data). Maturational parameters were not included in the survey. Thus a great deal of information relating to pre-adolescent growth and activity and its effect, if any, on adolescent development could not be obtained. Using a Bayesian approach with a suitable function these data would ultimately become available and could be enhanced by a single return visit to measure mature heights at a later date. Another study in Mexico City (Faulhaber, 1982, personal communication) examined adolescents in a mixed longitudinal study. The adolescents were in two groups followed for three years from ages 9 to 11 and 11 to 13 years. The study terminated because of shortage of funds but would provide useful descriptive information on adolescence if a Bayesian approach were used in the analysis.

OTHER GROWTH FUNCTIONS

The relationship of the Bayesian approach to growth functions or models in general must lie with the number of parameters required to fit the function and how well these may be determined prior to the iterative process of fitting.

Eight main functions have been described containing from 3 parameters in the case of Count (1942) to 9 parameters in the case of Bock's triple-logistic (Bock & Thissen, 1976). Gompertz (1825) and Jenss-Bayley (1937) contain 4 parameters and the Preece-Baines Model 1, 5 parameters. Using an iterative routine the starting values for the Preece-Baines curve may be very general, say within 10% of actual, and still obtain convergence within few iterations for a relatively smooth dimension such as height. More iterations may be required for dimensions having a less formal growth pattern such as biacromial or bi-iliac diameters. The greater the number of parameters the more critical it is, in terms of computer time and goodness of fit, to start with accurate estimates of these parameters. The usual approach of entering "guestimates" or previously published estimates of population mean parameter values is considerably enhanced using the Bayesian approach in which information on the variance and covariance are also entered so that greater information is available to the fitting process. The choice of the fitted function is still in the hands of the analyst but the precarious job of estimating starting values for parameters is made more accurate.

The main problem with the techniques described by Bock, from a purely selfish viewpoint, is the almost incomprehensible mechanics of using it. Indeed much of the theory presented by Bock in previous papers requires a very solid and sophisticated mathematical background to understand it. If anything is guaranteed to reduce the use of a growth function, no matter how good it may be, it is its apparently complex appearance and the lack of readily available computer programs to implement it. For the Bayesian estimation to become an accepted and widely applied method of parameter estimation, which it certainly should be, readily available computer programs must appear in a comprehensible format.

Notwithstanding this last comment Bock's approach is invaluable. To my knowledge it is the first attempt at estimating complete growth curves from a minimum number of actual data points.

REFERENCES

Bayley, N. & Pinneau, S.R., 1952, Tables for predicting adult height from skeletal age: revised for use with Greulich-Pyle hand standards. J. Pediat. 40:423

Boas, F., 1892, The growth of children. Science, 19:256, 281; 20:351

Bock, R.D., & Thissen, D., 1976, Fitting multi-component models for growth in stature. Proc. 9th Biomet Conf, 1:431

Bock, R.D., & Thissen, D., 1980, Statistical problems of fitting individual growth curves, in: "Human Physical Growth and Maturation: Methodologies and Factors", F.E. Johnston, A.F. Roche, & C. Susanne, eds., New York: Plenum p265

Count, E.W., 1942, A quantitive analysis of growth in certain human skull dimensions. Hum. Biol. 14: 143

Goldstein, H., 1979, "The Design and Analysis of Longitudinal Studies" London, Academic Press.

Gompertz, B., 1825, On the nature of the function expressive of the law of human mortality. Phil. Trans, Royal. Soc. 115: 513

Jenss, R.M., & Bayley, N., 1937, A mathematical model for studying the growth of a child. Hum. Biol. 9:556

Johnston, F.E., Roche, A.F., & Susanne, C., 1980 "Human Physical Growth and Maturation: Methodologies and Factors". New York, Plenum.

Karlberg, P., Taranger, J., Engström, I., Karlberg, J., Landström, T., Lichtenstein, H., Lindström, B., & Svenberg-Redegren, I., 1976, Physical growth from birth to 16 years and longitudinal outcome of the study during the same period. In: The Somatic Development of Children in a Swedish Urban Community. Acta Paed. Scand. Suppl. 258:7

Kendall, M.G., & Buckland, W.R., 1971, "A Dictionary of Statistical Terms", 3rd ed., London, Longman.

Patterson, H.D., 1950, Sampling on successive occasions with partial replacement of units. J. Roy. Stat. Soc B 12:241

Preece, M.A., & Baines, M.J., 1978, A new family of mathematical models describing the human growth curve. Ann. Hum. Biol. 5:1

Roche, A.F., Wainer, H., & Thissen, D., 1975, The RWT method for the prediction of adult stature. Pediat. 56:957

Tanner, J.M., Hughes, P.C.R., & Whitehouse, R.H., 1981, Radiographically determined widths of bone, muscle and fat in the upper arm and calf from age 3-18 years. Ann. Hum. Biol. 8:495

Tanner, J.M., Whitehouse, R.H., & Takaishi, M., 1966, Standards
 from birth to maturity for height, weight, height velocity and
 weight velocity: British children 1965. Arch. Dis.Childh 41:
 454, 613

DISCUSSION REPORT

(Methodology of Height Prediction)

Ole Lammert

Institute of Physical Education
Odense University
Denmark

The following took part in this discussion: B. Bogin, R.D. Bock, N. Cameron, A. Ferro-Luzzi, L. Greco, R. Hauspie, T. Hoang, F. Johnston, O. Lammert, B. Malina, W. Mueller, O. Neyzi, A. Roche, M. Roede, J. Tanner, and J. Van Wieringen.

Recently two reviews (Bock and Thissen, 1980; Preece and Heinrich, 1981) on prediction methods were published. Some of the points mentioned in this discussion were dealt with in detail in these reviews.

Summary of R.D. Bock's paper

Professor Bock presented a mathematical model which can predict the height of a subject with an accuracy of about ±3-5 cm. His formula is based on separation of the normal growth curve into three different parts. By the use of the triple logistic method he arrives to a formula which can, in principle, predict final height from a single measurement of age and height, but greater precision is attained when several measurements are used.

He emphasized that the model is specifically useful when applied around the ages of 10 for girls and 12 for boys. These are the age groups which pediatricians are most likely to be presented with, when the child and/or its parents gets worried about a rather short stature.

Summary of T. Hoang's comments

Professor Hoang congratulated Darrell Bock on his new contribution to this methodological area. She went on to discuss the accuracy of prediction of mature height. She found that Prof. Bock's model seems to disregard the correlation between mature stature and stature measured at previous ages, even at the age of 2 years.

She then asked if the "pulling-in effect" observed in a typical child had any influence on the normal prior measurement.

Finally, she commented on the feasibility of prediction from cross-sectional standards where secular changes of height are present. Such data are influenced by time, age, and the fact that the subjects in the different age groups are thus taken from different birth cohorts. Prediction from cross-sectional data therefore involves three things:

1. Determination of the percentile into which the individual's height falls.
2. The possibility of analyzing the pth percentile curves of the past and the changing population distribution of estimated heights.
3. The assumption that the individual's height track remains in the same interval of the population distribution.

She asked if such a scheme of prediction is feasible?

Summary of N. Cameron's comments

Dr. Cameron commented on Prof. Bock's technique in terms of its accessibility from the user's point of view and its accuracy when compared with other available growth predicting models.

He described the different results one gets when using different methods for prediction and pointed out that the present model requires very few data.

Dr. Cameron praised the usefulness of the model, because it lets one add later measurements after the first prediction and thus obtain more reliable predictions.

In connection with longitudinal data where values are missing, this model has the advantage that the information from all subjects can be used to improve the estimates for each individual.

He also mentioned the model's usefulness when one is working with longitudinal data where, for example, no measurements are obtained during adolescence; here it may be useful to use, for example, peak height velocity as a centering point.

Finally, he pointed out that from the user's point of view it is a weakness that the way the model works a very solid and sophisticated mathematical background is needed if one is to understand how the model operates.

DISCUSSION

Comparison with other predicting models

The present method require knowledge about the general growth pattern and mean heights of the population for which it is to be used. Furthermore it needs in principle only one, but in practice preferably several, measured observations of the individual for the prediction of the final height. These observations need not necessarily be spaced exactly one year apart as other methods require. Thus, unlike other methods, no interpolation is necessary with the use of this method. The method will in principle predict any kind of growth as long as the general pattern of growth is known, but at present it has only been applied to height.

Furthermore, this method predicts the growth pattern of the individual from the population pattern and the known standard errors. It is possible to follow a child along its percentile, and also observe when illness or treatment have effects on the child's growth. This is an advantage. The precision is improved when previous measurements of height are known, as for example from school records that are obtained regularly in the Scandinavian countries.

In the discussion the possibility of a further improvement of the method by the use of other observations, in particular bone age and midparents' stature, was debated. The fact that bone age determinations are more inaccurate in relation to time than chronological age was stressed, but the method itself can be used on bone age measurements. The use of midparents' stature as a supporting parameter was found to be a possible way to increase the accuracy of prediction in the future.

The present method can also be used for the determination of growth events not measured. In some studies the taking off point has been used, and many studies concentrated on the middle growth spurt or the period when the peak height velocity is observed. These events can be predicted from measurements obtained earlier or later in life. The fact that the method estimates the final height means that it can also be used in cross-sectional studies. This is useful in a survey where measurements are done on school-going subjects, and where only the present pattern of height distribution up to nearly the period when the final height is attained is measured. The Preece-Baines method can also be used for this kind of prediction.

An interesting question with models for the prediction of growth, and in particular with this model, is their relationship to the physiology of growth. Unfortunately at present a lack of sufficient knowledge about the different physiological interactions with growth makes it impossible to relate the mathematical models for prediction of height to biology of growth. The triple logistic curve is argued (Bock, in his present paper) to be a model based on the idea that growth from infancy to maturity reflects three cycles of hormonal influence. At present longitudinal data of changes in the hormonal concentrations in blood or urine are scarce and incomplete. Research on this problem is said to be underway, both in London (Preece) and in Turkey (Neyzi). One of the problems associated with this kind of longitudinal survey is the fact that the diurnal level of the hormones varies. At present one obtains one blood sample with a hormone level relevant for only that particular moment, but what really is needed is a mean sample taken over <u>at least</u> 24 hours consisting of subsamples obtained at regular and not too widely separated intervals. This is not possible at present. Furthermore, the standard deviation of these measurements is much greater than those of height or weight measurements. This increases the problems when such data are used for prediction purposes as well as when the models are to be used as models of biological events related to growth. Excellent reviews of these endocrinological growth relations were presented by Prof. Prader in his lecture "Biomedical and endocrinological aspects of normal growth and development" and by Dr. Preece in his review "Endocrinology of normal puberty" at the 3rd International Congress of Auxology.

<u>Discussion of the influence, if any, of the fact that the method was made to fit the Fels data</u>

The discussion went further into the fact that the method was built up to fit the Berkeley Growth Study data using the average values of the growth parameters for boys and girls taken from Tuddenham and Snyder (1954). The method thus derived was used on the Fels data to see how accurate the predictions were as well as which cases this method was unable to predict. The question to what extent these samples could be argued to be equal to a normal distribution of genetic and environmental mixtures was discussed. It was first argued that the number of variables which ought to be looked at for a determination of the degree of environmental homogeneity was unknown, and second that sufficient knowledge about the homogeneity, and thus the influence of socioeconomic factors, was also lacking. The tendency to think of the different growth study cohorts as homogeneous, in the face of such uncertainty concerning the overall environmental influence, should therefore be viewed with caution.

The possibility of using this method for prediction on a quite different sample from, for example, Mexico was discussed. Prof. Bock stated that either you can use the method to calculate new sets of standards for the investigated ethnic sample and after comparison with the ethnic data you again can repeat this, thus finally ending up with proper Mexican standards, or if you have these standards already, you can straightaway use them in the model.

The problem of to what extent the secular trend could be accounted for was brought forward, and it was suggested that it could be included if assumptions were made as to the magnitude of the secular trend at present. This could be based on comparisons of the three generations of the Fels data and the assumption of a certain constancy in this phenomenon, or the presumed trend could be deduced from measurements early in childhood. Such considerations should be viewed with caution because the secular trend was found to be present in early childhood as well as after the age of 12 years. For the mathematical computation, only the upward drift of the curves should be taken into account, because the growth curves in themselves are fairly constant. Anyway, one always has to go back to data in the population for comparison, regardless of the method one uses.

Limitations of the use of predicting models

The limitations of this method include, as with other methods, the inability to distinguish the hard-to-predict cases where either genetic or environmental defects are hidden and first become visible late in the growth period.

These hard-to-predict cases in the Fells population were also found hard to predict by previously published methods. The fact that these hard-to-predict cases do occur, for at present unknown reasons, is a serious weakness in the use of these height prediction methods. For example, if one is investigating longitudinally a group of children participating in heavy exercise and wants to investigate if final height is influenced by this, one may wish to predict final height from measurements of age versus height, measured before and at the start of the participation in the training program. Later, one could then compare the predicted results with the actual pattern of the child's growth performance. If one gets a significant difference in final height, one cannot distinguish whether this is a hard-to-predict case or the result of an unusual environment.

Another limitation is related to the fact that any general changes in the population means must be apparent and included in the model during the calculations. Such changes could, for example, be the secular trend or significant changes in environmental conditions such as famine or disease in a certain population.

Prof. Bock's ideas about short-term measurements of fixed markers were argued to be unreliable because the process of growth was not a constant ongoing process with a fixed rate. On the other hand, if the measurements are repeated rather often over a period of time, this problem would perhaps be overcome.

Discussion of the use of the triple logistic method

The present model for prediction was recently compared with previous models in some details by Preece and Heinrich (1981). The discussion was therefore mainly related to the use of the calculated autocorrelation function, which made it possible to obtain measurements at irregularly spaced occasions. The researcher is thus given the freedom of not being forced to perform repeated measurements on fixed schedules. This was found to be an advantage, although it was stated that traditionally measurements in well-run studies were obtained at fixed intervals. During the discussion it was stressed that the present method can be used with an ordinary microcomputer. At present the software is not available, but it will presumably be ready in the early autumn of 1983. The computer language is Fortran. A new development in this field was announced. It is a nonlinear method completely free of any parametrization. It is called the Projection Pursuit Method, and is a method where the computer searches in a family of smooth curves until it finds the curve which fits the observed points. It was stated that at Stanford University work was in progress on this new method.

Literature

Bock, D., and Thissen, D., 1980. Statistical problems of fitting individual growth curves; pp. 265-290, in "Human physical growth and maturation: Methodologies and factors." Johnston, F.E., Roche, A.F., and Susanne, C., ed. Plenum Press, New York.

Prader, A. 1982. Biomedical and endocrinological aspects of normal growth and development, in "Proceedings of the 3rd Int. Congr. of Auxology," Belgium (to be published).

Preece, M.A., and Baines, M.J., 1978. A new family of mathematical models describing the human growth curve. Annals of Human Biology 5, pp. 1-24.

Preece, M.A., and Heinrich, I., 1981. Mathematical modelling of individual growth curves. Brit. Med. Bull. vol. 37, no. 3, pp. 247-253.

Preece, M.A., 1982. Endocrinology of normal puberty, in "Proceedings of the 3rd Int. Congr. of Auxology," Belgium (to be published).

Tuddenham, R.D., and Snyder, M.M., 1954. Physical growth of California boys and girls from birth to 18 years. Univ. of California Publications in Child Development 1, pp. 183-364.

Anthropometric Surveys

METHODOLOGY OF A CROSS-SECTIONAL NATION-WIDE GROWTH STUDY

J.C. van Wieringen and Machteld J. Roede

University Children's Hospital
"Het Wilhelmina Kinderziekenhuis"
Utrecht, The Netherlands

Introduction

By now, it has been well established that growth standards are an essential tool in the study of growth. National growth standards serve a multiple purpose. First of all, they are essential in curative medicine, as well as in child health and school health services, in order to evaluate the growth of individual children. They are not only useful in tracking down in time deviations from the normal growth pattern, but they also may offer a quantitative indication to give the parents the reassuring information that their child's growth follows an acceptable pattern.

Secondly, when growth measurements are collected at sufficiently large intervals of time and in adequate numbers of individuals, they may reveal secular shifts in the growth patterns of the population involved. It is generally accepted to consider such secular shifts as relevant indicators of the health of this population. Van Wieringen discussed before (1979) the undulating character of such secular shifts.

Moreover, national growth standards offer relevant information for industrial designers and architects.

As a consequence, it is certainly desirable that the auxologists in each country have up to date national growth standards at their disposal. The need was felt to summarize once more the criteria for acceptable national standards. This will be done on the basis of information on the Dutch nation-wide growth studies from 1965 (van Wieringen, 1972) and 1980 (Roede and van Wieringen, in press).

Motivation of the 1980 Dutch survey

In The Netherlands national growth standards have been created, based on nation-wide surveys of 1955 and 1965 (van Wieringen, 1972). However, the 1965 diagrams needed revision: nation-wide measurements on 8.5-year-old schoolchildren, repeated every three years (de Wijn, 1975), and yearly information on the height of Dutch conscripts indicated that the 1965 standards are no longer representative for the recent population of growing children. Consequently, a third nation-wide survey has been carried out to produce up-to-date growth references and to analyze the latest secular shifts.

Sample Design

To prevent the risk that secular changes steal into the material, data collection should be limited to a period as short as possible. For this purpose a cross-sectional design prevails over (mixed-)longitudinal studies.

Age Range

If possible, the sample should cover the whole age range at which growth takes place. That means: with measurements starting as early as possible after birth and continuing up to some years after the age at which the increase in body height really has stopped. As a consequence, in developing countries individuals up to an age of about 25 years have to be included in the sample. However, the experience is that it is difficult to obtain measurements on a sufficiently large, not biased sample of individuals who no longer visit school. The 1965 nation-wide Dutch survey has indicated that in The Netherlands in males the final adult height is reached around the age of 19 years. Thus, the 1980 nation-wide Dutch survey covers individuals up to 20.0 years of age. If only older adolescents still attending school are included, an over-representation of non-promoted pupils and more intelligent and economically more privileged ones, may be the result. To reduce such a selection it is advisable to include in the sample also schoolleavers and employed adolescents.

Sample Size and Age Classes

It is recommendable to plan beforehand the actual number of individuals that have to be included in a nation-wide survey. First, it should be decided according to which interval of age the material will be analyzed. In the Dutch surveys the age-class intervals were adapted to the general growth pattern. During the first year of life, when growth velocity is still high, smaller age-classes are appropriate. Consequently, the data have been classified into age-groups of one week in the first, and of two weeks in the second quarter after birth, and of three weeks in the second half year of life; from the first year on age-classes of three months are sufficient. This has as a consequence that the total age-range of the 1980 survey, covering 20 years, includes 105 successive age-classes.

Within the age-class the growth variable increases. As a result, the dispersion of the calculated variable is somewhat greater than the dispersion at the mid-point. However, the experience of the 1965 survey was, that for the above described relatively small age intervals, a correction according to the formula given by Healy (1962) is not necessary.

Growth charts should not only visualize median values, but also the variability at successive ages. Therefore at least percentiles 10 and 90 should be drawn, and if possible 3 and 97. In order to compose percentiles, sufficient numbers per age-class are required. In the 1980 Dutch survey a sample was planned of at least 150 boys and 150 girls per age-class at infancy, of 300 boys and 300 girls for the 1- to 5-year-olds, and of 150 boys and 150 girls from 6-19 years of age. Table 1 shows the originally planned and the actually obtained numbers of individuals.

Table 1. Size of sample[+] of the nation-wide survey 1980

age in years	planned	actually obtained		
		total	males	females
0	7.800	8.301	4.279	4.022
1 - 4	9.600	7.848	4.110	3.738
5 - 9	7.200	8.797	4.447	4.350
10 - 14	6.000	9.107	4.547	4.560
15 - 19	6.000	7.818	4.197	3.621
total	36.600	41.871	21.580	20.291

+ healthy individuals of Dutch origin

Co-workers

The actual data collection can either be performed by one specialized team, or by various teams, spread all over the country. When one team is in charge of all the measurements, a survey-van can be used provided with standardized equipment. This has the advantage that all children are measured by the same observers, and with the same instruments, what may reduce interobserver errors. However, in view of the large numbers of individuals that have to be included in the study, and the limited period of time allowed for a cross-sectional survey, the overloaded sampling program may take too great demands on such a team.

In the Netherlands the second procedure for data collection has been used: the actual measuring and recording have been carried out by teams of health clinics for infants and toddlers, school health services, and vocational medicine (125, 52, and 5 teams, respectively). Data collection has been performed during their daily routine.

Co-working of different teams is only justified if they are well instructed. Accordingly, a written protocol with detailed instruction was sent to all co-working teams. Next, they were personally visited for strict instructions concerning methods of measuring and registration. During this visit scales and measuring-instruments were verified by checking them against standard weights and measuring-lath.

Methods

The methods of measuring and instruments were as described elsewhere (Cameron, 1978; van Wieringen, 1980).

Population sample

As yet, there is no conception of the same tenor about the choice of population on which growth standards have to be based. The International Union of Nutritional Sciences (1971) advises for each country the use of national growth standards. Although there are arguments against this statement, we are convinced that it is discouraging to compare the growth of individuals living in bad circumstances with the standards of another country which shows an advanced positive secular shift of the general growth pattern. For the same reason nationally representative growth diagrams are preferable to those only based on a sub-population which lives in a relatively well-off situation.

Basically, the sample design of the third Dutch survey was similar to the one of the previous studies in 1955 and 1965. Characteristic of these studies is the fact that the sample is commensurate with the actual population as far as occupational status and distribution over the country are concerned.

Socio-economic classes

In the Netherlands about 95% of infants and toddlers visit the health clinics very regularly, while almost all schoolchild-

ren are seen by school doctors at regular intervals. The
cooperating teams have been instructed either to measure all
children seen at a certain session, or, when only small numbers
were required, to measure at random part of the children, e.g.
every second or third child. The cooperation of numerous health
services garantees a nearly optimal representation of the
population in the sample as far as occupational status of father
is concerned (Table 2). The higher percentage of group 2 and the
lower percentage of group 3 for adolescents may represent the
career of the father but one has to take into account some bias
in the sampling in these older agegroups (see also Table 3).

Table 2. Percentage distribution according to age and occupational status of father
Boys   survey 1980

Age in years	occupational status father			
	1	2	3	4
0	8.0	28.9	52.6	10.5
1 - 4	7.4	30.4	51.9	10.2
5 - 11	7.5	32.5	52.4	7.6
12 - 16	7.2	37.2	46.6	8.9
17 - 19	8.5	37.1	44.2	10.1

1 = higher occupational status
2 = middle occupational status
3 = lower occupational status
4 = unknown and unemployed

Geographical distribution

Recent Dutch demographic data on the number of inhabitants
per province and the size of municipalities have been the starting point on which the number of cooperating teams have been
determinded. For instance, about 40 per cent of the sample had
to be collected in the closely populated coastal area, while
about three to four per cent sufficed for some of the relatively
sparsely populated provinces. The planned and the actually obtained samples have proved to differ hardly in this respect.
In order to meet the variation in number of inhabitants, data
have been collected in almost 200 places, ranging from small
villages to big cities.
Standardization of the sample of the adolescents according
to province, or to the distribution of father's occupational
status for schoolchildren had no effect on the mean values
(Table 3).

Table 3. Standardization of mean height (cm) of 19 years old males according to the distribution of father's occupational status of schoolchildren
The Netherlands survey 1980

occupational status	school children	18.5- 19.5 years old males	
	%	%	mean height
higher	7.25	10.1	184.56
middle	32.55	39.1	182.54
lower	52.40	38.2	182.26
rest	7.90	12.5	181.62
total	100.00	99.9	182.52

Mean according to standardization   182.47

Other countries may not have such a well organized chain of infant health centres and school health services that can be enlisted in the data collection. Yet, sampling covering the whole geographical area can be well achieved as has been proven by the 1972 Cuban national growth study (Jordan et al., 1975).

Variables

It is imperative to record for each subject exact information on sex, date of examination and birth date. The latter still is a source of uncertainty in many developing countries where birth notice is not yet so strictly arranged. In the 1980 survey also place of residence of the subject was recorded for analyses on geographical distribution. Information on the origin (father and mother both being Dutch, or not) enabled to separate afterwards autochtones from the immigrants. Beside these data, it is advisable to record family size and birth order, if possible supplemented by data about occupation of father and mother, and for children of school age, the type of school they are visiting.

Height. Height should be included in all studies on growth even though it is the sum of three parts of the body (head, trunk, legs) which each have a different type of growth pattern. The strong relation found in numerous populations between a secular increase in height and a decrease in morbidity and mortality has made height the most general and popular indicator of the health status of a population, as well as of a growing individual.

Weight. Also data on weight show secular changes, and it should certainly be included in any growth study. Weight for height is more appropriate for the medical and biological judgement than weight for age (van Wieringen 1978). It is worth trying to obtain data on a sufficient number of subjects to allow for weight for height analyses.

Head circumference. Most of the growth of the head takes place during the first year of life. Measuring of this variable during infancy may offer a reliable method to trace deviations in the growth of brains and/or skull at an early time.

Puberty rating. In the age range from 8-19 years evaluation of the development of sex characteristics gives an essential indication on the maturation at adolescence. Since most of these variables cannot be measured quantitatively the score is based on an estimation of the maturing stage. For the Dutch 1965 and 1980 surveys the physicians of school health services that were willing to cooperate on this special part of the survey got an extra instruction.

Recording

When data for a nation-wide survey are being made by one single team traveling all over the country with one standardized set of equipment, recording can be made semi-automatic. Then, it is desirable that the measurement results can be displayed on the actual moment of measuring to enable the observer to check its valdity. The Dutch data, collected by numerous observers, were all recorded on special recording forms that were sent back to the Children's Hospital in Utrecht.

Data processing

In the 1980 survey about 46,000 forms have been returned in the course of 1979-1980. The data have been carefully checked on misreadings and other imperfections, encoded, and put on file. After rejecting outliers, and setting apart data on children of non-Dutch origin (mainly of immigrant workers, n = 3440, and on prematures and children with pathology, n = 452), the data of a total number of 21,600 healthy Dutch boys and 20,300 healthy Dutch girls could be used to compose the new growth references.

Curve fitting

The nature of growth curves is not easy to describe by computer programs. Several methods are developed, but in the 1980 survey it was preferred to do part of the smoothing of the curves by eye.

All below applies to boys and girls separately.

For all 105 age-groups described above, the computer calculated for both height and weight, number of subjects, mean and standard deviations (s.d.) as well as the values of the percentiles 3, 10, 25, 50, 75, 90 and 97. As expected, comparison of the means and P50 values revealed a normal distribution of height except at adolescence and skewness of weight. Accordingly, two different methods have been applied for hand and eye smoothing of the curves.

Height. The computed s.d. values and P50 values of the successive age-groups have been plotted graphically, and eye smoothed. Next, these fitted curves were read on 104 age points. Based on these smoothed P50 and s.d. values, for the same age points percentiles values have been calculated according to the formula (van Wieringen, 1972):
P3, P97 = P50 -, + 1.881 x s.d., respectively
P10, P90 = P50 -, + 1.282 x s.d., respectively
P25, P75 = P50 -, + 0.675 x s.d., respectively

These percentiles have been plotted graphically. Next, on the same diagrams, the originally computed percentile values of the age-groups have been plotted also. Before puberty and towards adult age, the two series of the points proved to differ hardly. But in puberty they deviate due to the skewness of the adolescent growth spurt in the higher percentiles as compared to the lower ones. Final eye fitting by combining the visual impression of both series resulted in reading the required standard values at suitable age points. For smoothing the curves of the oldest adolescent males, the values of height of conscripts that have been examined in the same period were used as a point of reference.

Weight. Means of the computed median values per age-group were plotted graphically, eye fitted and then read at age points. The differences between the computed percentiles values (3, 10, 25, 75, 90, and 97) and these smoothed P50 values have been plotted at both sides of the P50, which was drawn as a straight horizontal line. This method is illustrated in Fig.1. These curves have been eye smoothed, and read. Next, these smoothed differences have been either distracted or added to the smoothed P50 values. These new values have been plotted together with the originally computed percentile values. The required standard values at suitable age points were obtained by final eye fitting combining the visual information of both series.

Weight for height. The large number of the Dutch sample permitted to calculate the distribution of weight for height. A preliminary computer analysis gave 10, 50, and 90 percentiles for weight for 2-cm height classes. After plotting, smoothing, and reading, age-independency over large age ranges became evident. Then final computer analyses have been made for infants and young toddlers, covering the height range of 47-99 cm, for $1.0 - \leqslant 13.0$, $13.0 - \leqslant 16.0$, and $16.0 - \leqslant 20.0$ years of age, respectively.

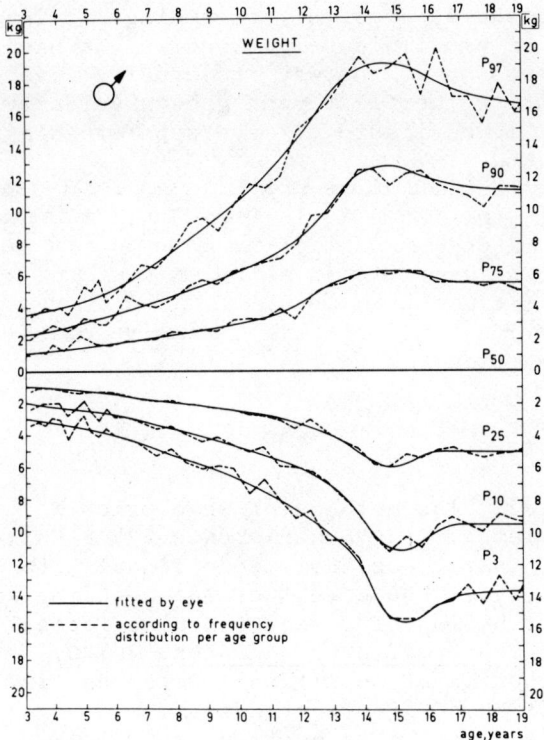

Fig.1. Method of graphical graduation of percentile curves. By drawing the median as a straight line, the often subtle changes in the distribution can be visualized.
van Wieringen, 1972

At last, all percentiles of height, weight, and weight for height were checked on regularity and graduality of the changes between the successive values.

<u>Sex characteristics</u>. The data obtained on sexual maturation should be classified in age-groups of three months. Sigmoid curves reflect the increasing proportion that has passed the beginning of a certain stage in relation to increasing age. By probit-analysis or the log-method the age-points can be calculated where 10, 50, and 90 per cent of the population has reached a certain stage of development.

Growth standards and diagrams

For practical use by physicians in curative medicine and in health centres, it is imperative to visualize the standard values in diagrams. Regularity or deviation of a child's growth pattern can only be detected if the child's measurements are plotted successively in a diagram. Tabulated measurements of a child's height and weight hardly give a suitable impression of the way it actually develops.

Since during the first year growth velocity decelerates extremely, it is advisable to have separate diagrams for infants.

In the Netherlands there are diagrams for 0-year-olds, for 1- to 6-year-olds, and for 3- to 19-year-olds. In the latter the development of sexual characteristics can also be recorded.

One should always keep in mind that diagrams are no more than an expedient, ard may never be used instead of an integral medical examination, nor are they strictly normative for what is abnormal, acceptable, or optimal.

References

Cameron,N., 1978, The methods of auxological anthropometry, in: "Human growth 2", F Falkner and J.M. Tanner,eds, Baillière Tindall, London.

Healy, M.J.R.,1962, The effect of age-grouping on the distribution of a measurement affected by growth. Amer.J. phys. Anthrop. 20:49.

International Union of Nutritional Sciences, 1971, The creation of growth standards: a committee report of a meeting in Tunis. Am. J. Clin. Nutr. 25:218-220.

Jordan,J,Ruben,M, Hernandez,J, Bebelagua, A, Tanner, J.M, and Goldstein,H., 1975, The 1972 Cuban national child growth study as an example of population health monitoring: design and methods, Ann Hum.Biol., 2:153-171.

Roede,M. and van Wieringen,J.C., in press, Growth diagrams 1980 Netherlands. Third nation-wide biometric survey on 0-19 year olds. T.Soc. Geneesk.

van Wieringen, 1972, Secular Changes of growth, 1964-1966 Height and weight surveys in The Netherlands in historical perspective, thesis, Leiden.

van Wieringen, 1979, Secular growth changes and environment.- An analysis of developments in The Netherlands, 1880-1978. Coll. Anthropol 3:35-47.

van Wieringen J.C.,1980, Surveys of physical growth and maturation, in: Human physical growth and maturation, methodologies and factors, F.E. Johnston, A.F. Roche and C. Susanne, eds., Plenum Press, New York.

de Wijn, J.F., 1975, Somatometrisch onderzoek, in: Zesde
    oriënterend onderzoek omtrent de voeding en voedings-
    toestand van 8-jarige schoolkinderen in Nederland
    (1973/1974) , Rijswijk.

THE USE OF ANTHROPOMETRIC SURVEYS FOR STUDIES

ON THE GENETICS OF GROWTH

              William H. Mueller

              The University of Texas
              School of Public Health
              Houston, Texas 77025 U.S.A.

    I will not attempt to review the entire area of human developmental genetics as Susanne (1980) has ably done this. Instead, I will try to review the most recent studies on the genetics of growth. In doing so, I will refer to studies on the quantitive genetics of adult stature, for example, as well as data which focus on size of preadults.

    The study of the genetics of growth is severely limited by methodological problems. The first of these is that almost all such studies involve measurements such as stature, bone age or tooth eruption. Measurements by necessity imply the use of quantitative genetic analysis. This method remains controversial, because it is based on coefficients of familial resemblance. It is well to remember that one of the characteristics of a genetic trait is its <u>segregation</u> in families. Segregation implies <u>dissimilarity</u>, and this dissimilarity is thus as much a mark of genetic influence as is similarity. Therefore much of the past decade's research on developmental genetics has been directed towards separating the contribution of shared environments and shared genes in familial likeness.

    The second methodological problem, related to the first, is that the human biochemistry of growth is not understood. While biochemical genetics has prospered since the 1955 discovery of the technique of electrophoresis, the characters studied and known best, are those of the red blood cell. Many of these are polymorphic, but few of them seem to be directly relevant for the growth process.

It would be constructive to think about possible approaches to the genetics of growth through the use of such genetic markers. Two suggestions involve correlation of gene markers and anthropometry. In the first, one searches for a polymorphic gene marker which might relate functionally to growth. An example is the group specific protein, a constituent of serum linked to vitamin D transport. There are apparent functional differences between individuals carrying alternate alleles of this genetic locus. The possibility that these differences could be related to differences in rates of bone growth, is intriguing. To my knowledge genetic marker laboratories have not been a standard part of the growth institutes. The biology of growth could be greatly enhanced by the inclusion of such facilities in studies of human growth.

A second possibility would correlate growth with degree of heterozygosity of the individual. Here the loci used need not be related to the growth process, but would simply be reflective of the overall heterozygosity of the individual. Studies on the relation between morphology and hereozygosity have been carried out in non-human organisms for many years. These studies have shown that the morphologic mean is rarely related to degree of heterozygosity, but that variation in growth is so related. With respect to variation there are two hypothetical possibilities :
1) If Heterozygous individuals are less variable than homozygotes, this would suggest that heterozygosity leads to greater stability or canalization of development. 2) If on the other hand, heterozygotes are more variable, this would suggest that heterozygosity leads to a greater ability to adjust to environmental circumstances, in other words greater developmental plasticity. Heterozygosity is simply defined as the number of sampled genetic loci at which an individual is heterozygous divided by the total number of loci sampled. Imlicit in this procedure is the sampling of a large number of polymorphic gene loci.

We are able to sample not a large number (4 highly polymorphic loci) in an Andean population in order to examine this question (Mueller et al., 1981). Our results while necessarily tenetative due to the small number of genetic loci involved, are presented in Table 1. The anthropometric shown is stature, which was adjusted first for age and sex, before searching for mean and variance differences between arbitrarily defined heterozygosity classes. Indeed the means did not differ significantly between these classes. However, we found that individuals who were more heterozygous, exhibited a _larger_ morphologic variance than homozygotes. This is presummed support for the second of the two hypotheses ennuciated earlier. This effect was evident in preadults as well as adults in our sample, but was statistically significant in adults only. The population studied could be described as being

Table 1

GROWTH AND HETEROZYGOSITY IN AN ANDEAN
POPULATION (STATURE)
(Mueller, Ferrel and Schull, 1981)

	PRE-ADULTS			ADULTS		
	N	Mean	Variance	N	Mean	Variance
HOMOZYGOTES	303	0.19	31.4	465	-0.12	24.6 **
HETEROZYGOTES	297	-0.12	36.8	447	0.18	30.7

** Between group difference significant at $p = 0.01$

under considerable environmental stress due to hypoxia and high rates of infections illnesses. Thus results may be relevant only to populations living under similar circumstances. The data are presented as an example of a new avenue of approach to the human genetics of growth.

Returning now to recent efforts in the study of the quantitive genetics of growth and development, it is probably true that the main effort in this field in the past decade has been investigation of the environment itself as a factor in the correlation between relatives. Two main areas of research are evident. The first calculates the genetic heritability of growth in non-western or less developed societies. The rationale of this approach is that the more stressful environments might alter individual genetic potential and result in parental stature for example, being less predictive of a child's growth than would be the case in societies in which the genetic potential may be maximized by optimal environmental circumstances. A second approach has attempted to investigate the role of environmental covariation in familial resemblance for physical characteristics.

Some results of the first approach are shown in Table 2. Shown are the average parent-offspring correlations for stature and weight in European populations (Mueller, 1976). The European samples come mainly from European and North American Growth institutes. The non-European samples were drawn from the developing countries or from poorer sectors of industrialized countries. The data refer to attained size at about age nine years, as my own research at the time focused on school aged children from the subsistance farming milieu in Colombia (South America). The average correlation among

Table 2

VARIATION IN PARENT OFFSPRING CORRELATIONS IN
STATURE AND WEIGHT WORLDWIDE (School Age Children)
(Mueller, 1976)

	Stature	Weight	N's (Approx.) Stature	Weight
EUROPEAN	0.37	0.31	1000	150
NON-EUROPEAN	0.29	0.25	2600	500

the European samples (r = 0.37) is higher than that among the non-European samples (r = 0.29), but the magnitude of the difference is small. Even if we were to consider the magnitude of the difference as meaningful, the reasons for reduced correlation in the non-Europeans could be unrelated to their environmental circumstances : Assortative mating for body size is less common among non-European groups, and age determination of children in some societies can be a problem. In our Colombian samples, with inclusion of children only for whom we could locate a birth certificate, the parent-offspring and sibling correlations in attained size and body composition were very similar to those reported earlier for industrial countries (Mueller and Titcomb, 1977, Mueller, 1977). This result has been replicated in Guatemala by Martorell et al. (1977). Reduced heritability of growth may be evident only among the very young as Russell (1976) found for Guatemalan pre-school children. All in all, it may be said that genetic influences on growth variability within a population seem little affected by adverse environments. However growth differences between social classes in developing countries are clearly more reflective of the environment children are experiencing than the genes they possess.

Turning now to the study of environmental covariation of relatives in physical growth. The approaches include : 1) The use of multiple kinds of relationships from which may be deduced environmental factors such as maternal effects or cross-generational environmental changes. This gets away from the reliance on estimates of genetic influence derived from single types of relationship such as the parent-offspring correlation by itself. 2) Another approach is the comparison of related individuals living in the same and different environments, or unrelated persons living in the same environment. 3) Thirdly we have the construction of environmental

Table 3

HERITABILITIES OF STATURE-MULTIPLE RELATIONSHIPS

	Parent-Offspring	Midparent-Offspring	Full Sibs	Half Sibs
Mat	.85	.65	.81	Mat 1.03
Pat	.61			Pat .56

(Roberts et al., 1978-West Africa)

	.77	.65	.76	--

(Furusho, 1961-Kyushu, Japan)

indices for use in partial correlation methods including path analysis. 4) Lastly we have investigations of the effect of demographic factors, most notably the age difference between pairs of relatives, to check for the effects of secular environmental changes.

An excellent example of the first approach is the study of the heritability of stature in a West African Population by Roberts et al. (1978). Here the culture favors polygamy, thus producing a large number of half-siblings as well as other more conventional relationships. Roberts' heritability estimates are shown in Table 3. These are compared to results of Furusho's (1961) study in a rural Japanese Population.

First we may note the wide variation in heritability of adult stature by type of relationship in these studies. This contrasts very sharply with the almost identical heritabilities of stature obtained in both samples. Of note are the generally high estimates of heritability of stature from the full sibling relationship, and in the West African sample, heritabilities derived through the mother-offspring relationship are much higher than those from male relatives. This is presumed evidence for a maternal effect. This did not occur in Japan, however Roberts et al. (1978) note that the environment has changed little over two generations in West Africa, while there is strong evidence for secular changes in the environment of rural Japan, thus possibly obliterating maternal effects in the latter sample. Unfortunately the half-sibling relationship is not available in the Japanese study, as this most clearly illustrates the extra effect of maternal body size on growth of offspring. These examples highlight the stability of

Table 4

SIBLING CORRELATIONS IN ANTHROPOMETRIC GROWTH
INDICES AT BIRTH TWO TAIWANESE EXPERIMENTAL GROUPS
(The Bacon Chow Study, Mueller and Pollitt, 1982)
ANTHROPOMETRIC INDEX (BRIEND, '80)

	N	1) Size	2) Stunting	3) Wasting
CONTROLS	110	0.46**	0.46**	0.33**
Experimental[a]	109	0.44**	0.16*	0.02
Chi Square		0.03	6.09*	5.55*

* p = 0.05   ** p = 0.01

[a] In Experimental Group, Younger member of pair nutritionally supplemented in-utero

heritabilities of growth in populations of different cultural backgrounds and the variability of estimates of genetic influence obtained from different relationships.

We now turn to an example of the second methodology. When environmental differences cut across family members, familial correlations may be dramatically affected as in shown Table 4. Shown are full sibling correlations in three anthropometric indices derived from principal components following Briend (1980). The three indices are uncorrelated and the measurements were taken at birth. The data are from the Bacon Chow study of nutritional supplementation of rural Taiwanese women during pregnancy. Due to the untimely death of the principal investigator, Dr. Chow, these data were never fully analysed. Dr. Ernesto Pollitt, a colleague at the University of Texas School of Public Health, applied for and received the data set from the U.S. Agency for International Development, which had funded the original study. In this study two groups of mothers were randomly assigned to a control which received a placebo, and a supplement or experimental group which received a high calorie/protein supplement during the pregnancy of a later born child. During the pregnancy of an earlier child, no treatment was given, although both children's growth was subsequently measured. Thus we have a classical genetic study design : The control siblings (N = 110) would have experienced identical nutritional treatments; The experimental siblings, however, would have experienced very different nutritional circumstances in-utero.

One can see dramatic effects on sibling variation in the growth indices at birth (Table 4). While control siblings are correlated, coefficients in the experimental group are depressed. This is particularly the case for indices of relative growth, the stunting and wasting indices. Evidently there is a large environmental component to the physical resemblance of siblings at birth via nutritional status. At the Sogesta conference Dr. Ingeborg Brandt was helpful in bringing to our attention the importance of indices of relative growth of infants as the best indicators of nutritional status in-utero. If it were not for her, we would never have carried out this investigation of relative size. She advocated the use of the Rohrer Index at birth (Brandt, 1980), and if this index is used the same dramatic effects on sibling covariation are evident. These observations are all the more amazing as we could detect no effects of supplementation on the group means. Hence, the quantitative genetic method may be more sensitive in detecting developmental changes as a result of this kind of nutritional supplementation, than more conventional approaches.

Let us now consider jointly the last two approaches to the study of familial environment as a cause of likeness of relatives in physical characteristics.

Table 5

SIBLING CORRELATIONS IN BODY MEASUREMENTS
MUELLER : ADULT SIBLINGS BY AGE-DIFFERENCE CLASS
BOUCHARD : PRE-ADULT SIBLINGS, PARTIALS WITH SES CONSTANT
(Bouchard, 1980)

Measurements	AGE-DIFFERENCE		Bouchard	
	7 yrs.	7 yrs.	r_{Sib}	$r_{Sib.SES}$
HEIGHT	0.45**	0.55**	0.53**	0.34**
SITTING HT	0.34**	0.53**	0.45**	0.24**
WEIGHT	0.39**	0.23	0.49**	0.29**
ARM CIRCUM	0.45**	0.24	0.41**	0.29**

* p = 0.05 ** p = 0.01

Shown in the left two columns of Table 5 are correlations between adult siblings in body measurements. The data are from rural Colombia and siblings are classed into two age difference groups. As far as I know Furusho (1963) was the first to use this demographic device to explain variability in sibling and parent-offspring coefficients of correlation for stature. The reasoning goes that the larger the age difference between a pair of relatives, the greater the environmental difference between them. This assumes that the environment is changing over time, and of course this may be true for some populations more than others and will also depend on the age difference distribution in the sample. Rao et al. (1975) also investigated the effect of age difference of siblings on familial covariation in stature in Brazil, and found marked effects. In our sample those siblings furthest apart in age (here 7 years or greater) had diminished resemblances in soft tissue anthropometry such as weight and arm circumference. This was not true for height or other measures of boney development. This suggested that heritability of adult height might be more stable than that of body mass.

Bouchard (1980) compared this approach to another method. The method he uses in his Quebec sample, is to construct an index of common environment and partial this out of the sibling correlation for body measurements. Environmental indices of this kind have now been used widely in path analysis of the genetics of measurements (Rao et al.; 1975). Bock (1980) has ably discussed the use of such indices in the context of path analysis. I present this simpler approach detailed in the right two columns of Table 5, to illustrate the use of this kind of index. In Bouchard's (1980) sample the reductions in the sibling correlations are quite dramatic suggesting at face value a large environmental covariability in these 10 year old Quebec children.

Listed in Table 6 are the components of several indices of environment used by different investigators. There are methodological limitations in the use of environmental indices, one such limitation is, that it would be hard to specify the ideal set of variables to comprise the index, applicable to all populations at all times. In this figure for example, we see a variety of variables used by different investigators. Parental occupation and education as well as income are usual, but in societies such as the Colombian one, it is hard to get an estimate of income for example, and more indirect measures need to be used. A second problem as Bouchard (1980) has noted is that if there is genotype/environment correlation the index will tend to remove genetic as well as environmental causes of physical resemblance. A third limitation, is that development is a continuously changing process, and an index measure usually at one point in time, is unlikely to be sensitive enough. Some idea of the sensitivity of these indices is shown by the correlation of stature with the indices, also given in Table 6. Rao et al. (1975)

Table 6

INDICES OF ENVIRONMENT OR SES

Bouchard (1980) Quebec-10 year olds R = 0.53

    Parental Education and Income (4)
    Income spent on food (2)
    No. Persons in Household (1)

Rao et al. (1975) (Brazil-adults) (R = 0.19)

    Parental Literacy and Social Level (2)
    Local Density and Location (3)
    Pathology (1)

Mueller (1975) Colombia-School ages R = 0.18

    Housing size and Materials (4)
    Sanitation and Utilities (3)

Johnston et al. (1980) Mexico-Pre-School R = 0.42

    Parental Education (3)
    Food/Non-food Expenditures (2)
    Sanitation (3)
    (Another 18 variables-no contribution to growth)

R = Correlation of growth and index

report a meager 0.19 as did Mueller (1975). Bouchard's is the highest at 0.53, and Johnston et al. (1980) report a correlation of index and growth of 0.43 in pre-school Mexican children. Clearly such socio-economic indices have had variable success in explaining environmental causes of resemblance of relatives.

In conclusion there are still some areas in which quantitative genetic method may still bear fruit.

First, we know quite a bit concerning genetic effects on attained growth, especially stature, we know little of genetic effects on growth <u>velocity</u>. The new class of fitted curves developed by Preece and Baines (1978) and Largo et al. (1978), provide ways of estimating growth parameters other than distance. These could fruitfully be put to use in genetic studies.

Secondly, much is known concerning the inheritance of physical growth, but much less is understood about the genetics of other measurements important in understanding the overall growth picture. These include measurements of physiologic function (Mueller et al., 1980), physical performance and activity (Malina and Mueller, 1981), body composition and biologic age or maturity (Susanne, 1980).

Thirdly, the development of better environmental indices for use with sophisticated methods of analysis, such as path analysis of growth, is clearly a desirable goal. Because these techniques, while promising, are only as the indices of environmental variability we have at our disposal.

References

Bock, R.D. 1980 Quantitative genetic analysis of variation in adult stature : The contribution of path analysis, In : Human Physical Growth and Maturation, F.E. Johnston, A.F. Roche and C. Susanne, ed., Plenum Press, New York.

Bouchard, C. 1980 Transient environmental effects detected in sibling correlations, Ann. Hum. Biol., 7 : 89

Brandt, I. 1980 Postnatal growth of preterm and fumm-tern infants, In : Human Physical Growth and Maturation, F.E. Johnston, A.F. Roche and C. Susanne, eds., Plenum Press, New York.

Briend, A. 1980 Fetal stunting, fetal wasting and maternal nutritional status, In : Maternal Nutrition During Pregnancy and Lactation, H. Aebi and R. Whitehead, eds., Verlag Hans Huber, Bern.

Furusho, T. 1961 Genetic study on stature, Japan J. Hum. Genet., 6 : 78.

Furusho, T. 1963 On the factors affecting the sib correlation of stature, Japan J. Hum. Genet., 8 : 255.

Johnston, F.E., Newman, B., Cravioto, J., Delicardie, S. and Scholl, T. 1980 A factor analysis of correlates of nutritional status in Mexican children, birth to 3 years, In : Social and Biological Predictors of Nutritional Status, Physical Growth and Behavioral Development, L.S. Greene and F.E. Johnston, Eds., Academic Press, New York.

Largo, R.H. Gasser, Th., Prader, A., Stuetzle, W. and Huber, P.J. 1978 Analysis of the adolescent growth spurt using smoothing spline functions, Ann. Hum. Biol., 5 : 421.

Malina, R.M. and Mueller, W.H. 1981 Genetic and environmental influences on the strengh and motor performance of Philadelphia school children, Hum. Biol., 53 : 163.

Martorell, R., Yarbrough, C., Lechtig, A., Delgado, H., and Klein, R.E. 1977 Genetic-environmental interactions in physical growth, Acta Pediat. Scand., 66 : 579.

Mueller, W.H. 1975 Parent-Child and Sibling Correlations and Heritability of Body Measurements in a Rural Colombian Population, Ph.D. Dissertation, University of Texas, Austin.

Mueller, W.H. 1976 Parent-Child and Sibling Correlations for stature and weight among school aged children : A review of 24 studies, Hum. Biol., 48 : 379.

Mueller, W.H. 1977 Sibling correlations in growth and adult morphology in a rural Colombian population, Ann. Hum. Biol., 4 : 133.

Mueller, W.H. 1978 Transient environmental changes and age-limited genes as causes of variation in Sib-sib and parent-offspring correlations, Ann. Hum. Biol., 4 : 395.

Mueller, W.H. and Titcomb, M. 1977 Genetic and environmental determinants of growth of school-aged children in a rural Colombian population, Ann. Hum. Biol., 4 : 1.

Mueller, W.H. and Pollitt, E. 1982 Supplement effects on sibling covariation in anthropometry using multivariate growth indices, in preparation.

Mueller, W.H., Ferrell, R.E., Bertin, T., Barton, S.A., Rothhammer, F., and Schull, W.J. 1981 Genetic flexibility and morphological variation in populations residing at different altitudes in Chile, Am. J. Phys. Anthrop., 54 : 257.

Mueller, W.H., Chakraborty, R., Barton, S.A., Rothhammer, F., and Schull, W.J. 1980 Genes and epidemiology in anthropological adaptation studies : Familial correlations in lung function in populations residing at different altitudes in Chile, Med. Anthrop., 4 : 267.

Preece, M.A. and Baines, M.J. 1978 A new family of mathematical models describing the human growth curve, Ann. Hum. Biol., 5 : 1.

Rao, D.C., MaClean, C.J., Morton, N.E., and Yee, S., 1975 Analysis of family resemblance V. Height and weight in Northeastern Brazil, Am. J. Hum. Genet., 27 : 509.

Roberts, D.R., Biellewicz, W.Z., and McGregor, I.A. 1978 Heritability of stature in a west African population, Ann. Hum. Genet., 42 : 15.

Russell, M. 1976 Parent-child and sibling-sibling correlations of height and weight in a rural Guatemalan population of preschool children, Hum. Biol., 48 : 501.

Susanne, C. 1980 Developmental genetics of man, In : Human Physical Growth and Maturation, F.E. Johnston, A.F. Roche and C. Susanne, eds., Plenum Press, New York.

METHODS IN HUMAN GROWTH GENETICS

Charles Susanne

Free University Brussels
Laboratory of Anthropology
Pleinlaan 2, 1050 Brussels, Belgium

When we study growth and development, we are in fact interested in the variation of metric characters in a population. One component of this variation is attributable to genetic causes. When we are trying to situate the normality of a child or when we are making a prediction of his adult stature, for instance, we must take these genetic factors into account (Bock, 1982). These genetic influences depend on a model that Fisher developed in 1918 to reconcile biometry and the laws of Mendel.

The phenotypes (P) we are studying in growth and development are a function of the genotypes and of the environment. Using the formulation of Defrise (1981), we can propose

$$P = F(x_1, x_2, \ldots, x_p, z_1, z_2, \ldots, z_q) \tag{1}$$

where x_i are the genetic factors and z_i the environmental factors. This function can be analyzed as the sum of three elements

$$P = G(x_1, \ldots, x_p) + E(z_1, \ldots, z_q) + f(x_1 \ldots x_p, z_1 \ldots z_q) \tag{2}$$

where G is the genetic component, E the environmental component, and f an interaction between G and E resulting from the nonindependence of the interaction between nature and nurture during development. We can however find many situations where the genetic factors are not independent of the environmental factors. For instance, an intelligent child will be more stimulated by intellectual influences of families or of schools.

In most analyses, the term f is neglected, though in the analysis of growth we can not exclude the possibility that some specific differences in environment have some influence on some particular genotypes.

Thus we have

$$P = G + E \tag{3}$$

In this system, if we suppose linearity with

$$G = x_1 + x_2 + \ldots x_p \quad \text{and} \quad E = z_1 + z_2 + \ldots z_q,$$

if p and q are large, and if the effects x and z have the same level and are independent, the central limit theorem teaches us that P will be normally distributed.

From (3) we can derive

$$V_p = V_G + V_E + 2\,\text{cov}_{GE} \tag{4}$$

or, since the covariance is often neglected,

$$V_p = V_G + V_E \tag{5}$$

The effects of genetic and environmental factors are additive and are randomly associated. The genotypic values do not influence the environmental values.

The environmental variance includes all variation of nongenetic origin, and thus also errors of measurement. Generally speaking, environmental variance includes nutritional and hygienic factors, but also climatic and cultural factors and maternal effects (prenatal and postnatal).

The causes are surely multiple. In a rather artificial way, we could classify the environmental differences into differences between individuals of the same family, differences between families, differences between socioeconomic groups, and differences between geographical groups. The variation existing between families or social groups of the same population results in an increase of the covariance between parents and children and thus of the correlations expected between parents and offspring.

In other words, the resemblance between relatives has not only a genetic origin but also an environmental origin. A common environment is shared by relatives, and this common environment is also a function of the degree of relationship. The environmental variance can be divided into two components, V_{EC} contributing to the variance between means of families but not to the variance within families, and increasing the covariance between relatives, and the remainder V_{EW} contributing to the within-family component:

$$V_E = V_{EC} + V_{EW}$$

In conformity with the attitudes of his period (1918), Fisher neglected V_E: "An examination of the best figures for human measurements shows that there is little or no indication of nongenetic causes." Even though nowadays we are particularly interested in the influence of the environment (and our interaction with it), we must admit that we find it difficult to analyze V_E or to quantify environmental influences. Equation (5) can also be written as

$$V_p = V_A + V_D + V_i + V_{EC} + V_{EW} \tag{6}$$

where the genetic variance is broken down into the variance of purely additive factors (V_A), the variance of dominant factors (V_D), and the variance of epistasis (V_i) or of interaction between genes of different locus. V_A refers to the hypothesis that the genes have an additive influence. Only these additive effects determine a selection value. However, it is possible to show that these different variances are a function of the frequencies of genes (Jacquard, 1977; Defrise, 1981). In other words, the genetic variance is a function of the studied population.

The degree of resemblance provides the means of estimating the amount of additive variances. In the Hardy-Weinberg equilibrium, neglecting all nongenetic causes and epistatic factors, the covariance between relatives is composed of the additive and dominant components and thus can be used to estimate these causal components (Table 1). Epistatic factors (and effects of linkage contributing to the epistatic component) contribute only a small fraction of the covariance. Even in experimental breeding these influences are hardly detectable.

To estimate the relative importance of heredity and of environment, a partitioning of the variance is proposed. The variance due to genetic factors is calculated as a proportion of the total variance. This is called the heritability. V_G/V_p, or heritability in the broad sense, expresses the relative importance of the genotypes. This value is used in human populations, but has no practical importance for breeders, who use V_A/V_p, the heritability in the narrow sense, which determines the degree of resemblance between relatives and which indicates the chief response of the population to selection (Table 1).

But even for traits with good additivity, such as finger ridge count or even stature, environmental influences are not negligible. These traits do not develop outside the environment: we cannot conclude that phenotypic differences are exclusively genetically determined.

Heritability will depend on gene frequencies and on the interactions between genotype and environment. It is not a constant

Table 1. Covariances of Relatives (Contributions of Additive and Dominant Variances), Regression (b) or Correlation (r), Heritability ($h^2 = V_A/V_p$), Expected Common Environmental Variance (V_{EC}), and Coefficient of Relationship

Relationship	Covariance	b or r	h^2	V_{EC}	Coefficient
MZ twins	$cov_{MZ} = V_A + V_D$	$r = \dfrac{V_A + V_D}{V_p}$	$h^2 \leqslant r$	$+$	1
DZ twins, full sibs	$cov_S = \dfrac{1}{2}V_A + \dfrac{1}{4}V_D$	$r = \dfrac{\frac{1}{2}V_A + \frac{1}{4}V_D}{V_p}$	$h^2 \leqslant 2r$	$+$	$\dfrac{1}{2}$
Parent-offspring	$cov_{PO} = \dfrac{1}{2}V_A$	$b = \dfrac{1}{2}\dfrac{V_A}{V_p}$	$h^2 = 2b$	$\pm$	$\dfrac{1}{2}$
Midparent-offspring	$cov_{MPO} = \dfrac{1}{2}V_A$	$b = \dfrac{V_A}{V_p}$	$h^2 = b$	$\pm$	$\sqrt{\dfrac{1}{2}}$
Grandparent-offspring, half sib, uncle/aunt, nephew, niece	$cov_{GO} = \dfrac{1}{4}V_A$	$r = \dfrac{1}{4}\dfrac{V_A}{V_p}$	$h^2 = 4r$	$+$ (in HS)	$\dfrac{1}{4}$
Double first cousin	$cov_C = \dfrac{1}{4}V_A + \dfrac{1}{16}V_D$	$r = \dfrac{\frac{1}{4}V_A + \frac{1}{16}V_D}{V_p}$	$h^2 \leqslant 4r$	$-$	$\dfrac{1}{4}$
Offspring-great grandparent	$cov_{GG} = \dfrac{1}{8}V_A$	$b = \dfrac{1}{8}\dfrac{V_A}{V_p}$	$h^2 = 8b$	$-$	$\dfrac{1}{8}$

value for the studied trait, it is a populational value. Even an elementary analysis of the formula for heritability

$$h^2 = \frac{V_G}{V_P} = \frac{V_G}{V_G + V_E} \qquad (7)$$

shows that h^2 will vary as a function of the environmental variance: V_E being small in rather homogeneous populations, the heritability will be high, V_E being large in heterogeneous populations, the heritability will be low. It is evident that the heritability as defined in (7) describes also the environmental background of the studied population. It is the multiplicity of studies in different populations which will help us better to evaluate the heritability. If h^2 is always equal to 0, it implies only environmental influences, but if $h^2 > 0$, it implies that genetic influences are acting in some way.

The choice of the kind of relationship used to estimate the heritability often depends on the available data. Theoretically the most reliable data are the regression of offspring on midparent (if the variance of the two sexes is equal) or the regression of offspring on one parent (only on the father, if an important maternal effect is suspected). The full-sib correlation (and even the half-sib correlation, if they are reared together for a long period) is the least reliable, since the contribution of the common environment (V_{EC}) is large and moreover an eventual contribution of dominance variance is also expected.

Twin studies have often been proposed for the study of heritability. The most frequently used formulas, although not theoretically founded, were proposed by Holzinger (1929):

$$H_1 = \frac{\sigma^2_{DZ} - \sigma^2_{MZ}}{\sigma^2_{DZ}}$$

$$H_2 = \frac{r_{MZ} - r_{DZ}}{1 - r_{DZ}}$$

Improvements have been suggested, for example, by Falconer (1980)

$$h^2 = 2(r_{MZ} - r_{DZ})$$

by Nichols (1965)

$$H_3 = 2(r_{MZ} - r_{DZ})/r_{MZ} \doteq h^2/r_{MZ}$$

and by Vandenberg (1965)

$$F = \sigma^2_{DZ}/\sigma^2_{MZ} = 1/(1-H_1)$$

All these formulas are not equivalent. When they are applied to the same sample, quite different results can be obtained. But in these analyses the effect of the common familial environment is not totally eliminated if we admit that the similarity of environment of monozygotes and dizygotes is not equivalent. For many traits, psychological character, for instance, the environment of monozygotic twins is more similar than the environment of dizygotes. A more recent approach in twin studies gives an estimate of the total genetic values G_T from the mean squares of the intrapair differences (M_{WDZ} and M_{WMZ}) (Christian et al., 1974):

$$\hat{G}_T = M_{WDZ} - M_{WMZ}$$

The factor $\hat{G}_T$ includes not only $1/2\ \sigma^2_a$ but also a part of the variance of dominance ($3/4\ \sigma^2_d$) and a function of epistasis including the genotypical values and the frequencies of genes. This model of Christian however supposes the covariance between genotype and environment of monozygotic pairs to be equivalent to the covariance of dizygotic twins. To use the factor $\hat{G}_T$, the environmental variance and covariance of monozygotic and dizygotic twins must be equal. If this is not the case, another estimation of the genetic values can be used:

$$\hat{G}_{CT} = G_T + ((C_{MZ}+C_{DZ}) + 2(\sigma_{GE}-\sigma_{GE+}))$$

with C_{MZ} and C_{DZ} the covariance of the environmental effects inside the MZ or DZ pairs, and σ_{GE} and σ_{GE+} the covariance of genetic and environmental factors for MZ and DZ twins.

For genetic studies, assuming that the characters are multifactorial and normally distributed, the inheritance can also be described by correlation coefficients calculated on large samples of relatives. Coefficients of correlation between parent and child are not really coefficients of heritability, but in a specific population they are very useful in determining a decreasing order of genetic influences (Susanne, 1971, 1975, 1980).

The correlation, first developed by Fisher (1918) and later by Wright (1934) in a path coefficient analysis, integrates various components: it is calculated between the phenotypes of relatives, but it implies not only the genetic link between gametes of the two relatives but also, in each generation, the correlation between individual haploid genic value and the phenotype (Fig. 1).

The genetic correlation, which is the only constant value and which is frequently called the coefficient of relationship, is

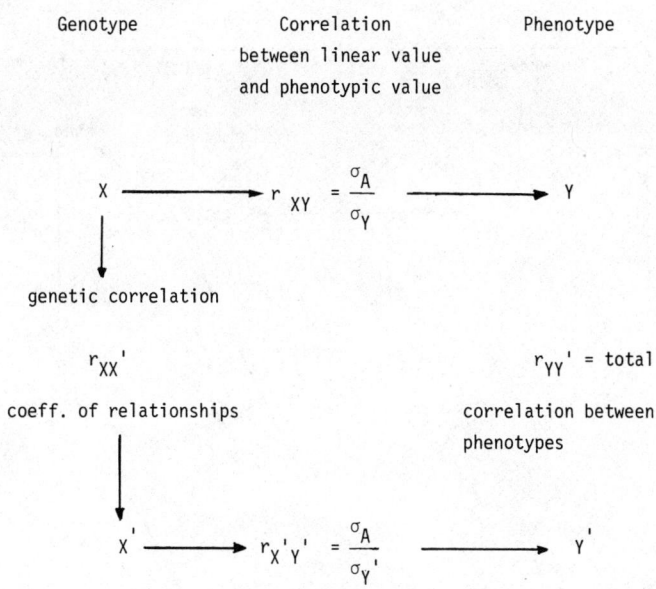

Fig. 1. Components of the correlation between phenotypes.

based on the probability that the relatives share identical genes (Table 1). To overcome the limitations of the model proposed by Fisher, Wright developed the method of path coefficients: the path diagram represents a model of linear relationship between the causal factors and the observed variables, in a scheme connecting zygotes and gametes; with suitable available data, parameters of the model can be estimated. All variables of the path analysis must be quantified and must have a continuous distribution. The path coefficient (P_{YX}) is the standardized regression coefficient between the causal variable and the response variable.

However, path analysis also neglects certain factors. For instance, epistasis is assumed to be absent. The full advantage of the method was developed by Rao et al. (1978), when they included the potentially important cultural sources of correlation. Cultural heritability is defined as the proportion of phenotypic variance due to family environment: V_E/V_P for children and $(V_E/V_P)y^2$ for adults (where y is a factor of differentiation between generations; if intergenerational differences are absent, y = 1).

In path analysis, the correlation between father and son is obtained by multiplication of a set of path cofficients (Fig. 2): the effect of genotype of the father on his phenotype (h_F), the coefficient of father-son relationship (½), and the effect of the

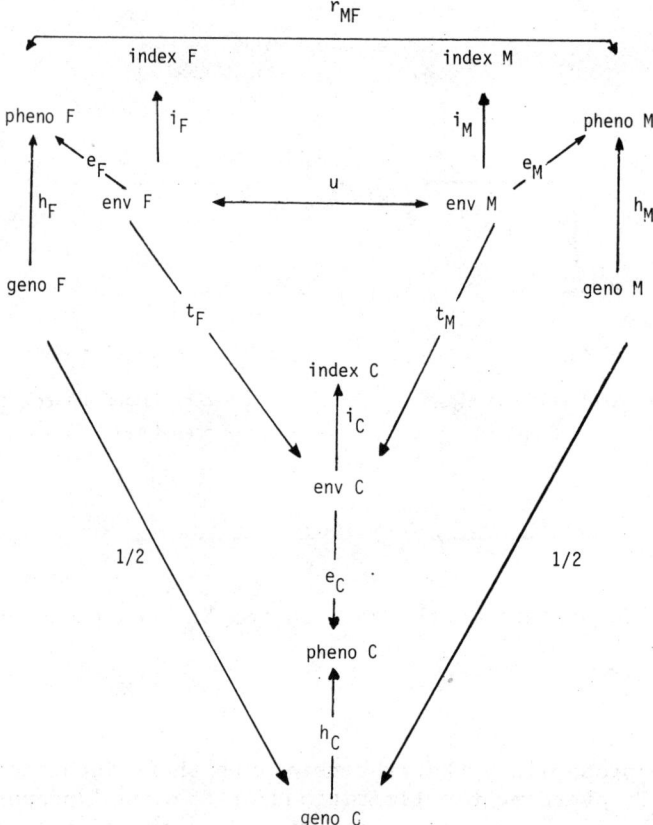

pheno = phenotype, geno = genotype, env = environment, index = environmental index, F = father, M = mother, C = child. The diagram has been simplified by assuming that there are no differences between sons and daughters.

Fig. 2. Path diagram for the parent-child relationship.

genotype of the son on his phenotype (h_C). To this genetic background the familial background is added: the effect of the father's environmental background on his phenotype (e_F), the correlation between the father's and the son's environmental background (t_F), and the effect of the son's environmental background on his phenotype (e_C):

$$r_{pheno\ FC} = h_F \cdot \frac{1}{2} \cdot h_C + e_F \cdot F_{FC} \cdot e_C$$

where

$$F_{FC} = t_F + t_M \cdot u$$

F_{FC} is the sum of the direct effect of the father's environmental background on his son (t_F) and the indirect effect resulting from the correlation of the father's and mother's environmental background (u).

Path analysis remains a linear model. Even with causes not directly measured, this linearity will give a good fit to the data. Path analysis differs from other linear models by the fact that it explains the interrelationship between the variables by a causal scheme. This technique has been increasingly used for instance in genetic epidemiology on cardiovascular risk analysis (Weinberg et al., 1979; Ward et al., 1980). It has the advantage of allowing tests of hypotheses, for instance, to test that genetic determination is equal to zero, or that sociocultural determination is zero, or that genetic and environmental determinations are equivalent. However, it requires that an environmental index be constructed. The analysis is valuable when a good quantification of the environment is possible, but one wonders whether in IQ studies an estimation by socio-economic status is sufficient (Rao et al., 1978). The nature of this analysis remains relative, as it was for the related Fisher methodology.

Many methods can be used (coefficients of correlation or regression, coefficients of heritability or path analysis), but the estimates are always relative to a specific population in its environment and time.

MULTIVARIATE ANALYSIS

Different techniques are also available on the multivariate level. They mostly consist in an extraction of factors to which the techniques used for individual measurements can be applied, such as the linear additive model or the path coefficient, or, for instance, the Christian model for twins.

Factor analysis can be applied to a group of measurements and loadings of the extracted factors; the measurements can be analyzed in the classic manner and the individual factor scores used for a sample of twins to measure heritability. An example on 11 face measurements is given in Table 2 (Defrise, 1982). It is evident that in this sample the genetic component is high for the individual measurements and also high (not higher, however) for the individual factor scores.

On psychological data, a multivariate approach to twin data has been proposed by Vandenberg (1956) and Bock et al. (1968) where only the within-pair covariance matrices are considered.

Table 2. Univariate Analysis and Factor Analysis of 11 Face Measurements in a Twin Sample (Twins ♂♂ 47 MZ and 35 DZ).

	Holzinger heritability	P of G_{WT} Christian (model)
(A) Univariate analysis ♂		
1. frontal breadth	.70	.000312
2. bizygomatic breadth	.69	.001687
3. bigonial breadth	.78	.000001
4. physionomic height	.79	.000019
5. facial height	.63	.008527
6. nasion-stomion height	.42	.003399
7. nose height	.37	.004304
8. nose breadth	.55	.000058
9. internal ocular breadth	.60	.009514
10. height of lips	.80	.000004
11. breadth of mouth	.72	.096704
(B) Factor analysis		
Factor 1 NR* height 4,5,6,7	.70	.000085
VAR height 4,5,6,7	.39	.032479
Factor 2 NR breadth 1,2,3,8,9,11	.73	.004092
VAR breadth 1,2,3,11	.79	.001884
Factor 3 NR 10,8,9	.64	.000006
VAR 10,8,9	.56	.003765
Factor 4 NR 8	.69	.002525
VAR 4,10	.79	.000000

*NR = principal components without rotation
VAR = varimax

Component pair analysis was developed by Iagolnitzer (1977) and was applied to twin dermatoglyphics (Iagolnitzer, 1978) and to mineral metabolism (Darlu, 1981). It may be regarded as a regular component analysis performed on paired data arranged in a four-block array, where within and between matrices are considered. The matrices are constructed of the elements a_{ij}, the individual differences from the mean value for the first twin, and b_{ij}, the individual differences from the mean value for the second twin; i = 1, ..., n is the number of twin pairs and $j = 1, \ldots, p$ the number of measurements:

$$A_{nxp} = \begin{vmatrix} a_{11} & \cdots & a_{1p} \\ \cdot & a_{ij} & \cdot \\ a_{n1} & \cdots & a_{np} \end{vmatrix} \qquad B_{nxp} = \begin{vmatrix} b_{11} & \cdots & b_{1p} \\ \cdot & b_{ij} & \cdot \\ b_{n1} & \cdots & b_{np} \end{vmatrix}$$

A principal component analysis is performed on the covariance matrix in a 2p space, grouping sums and differences of A and B:

$$\text{cov } T_{sd} = \frac{1}{2n} \begin{vmatrix} A'+B' & A'+B' \\ A'-B' & -(A'-B')' \end{vmatrix} \begin{vmatrix} A+B & A-B \\ A+B & -(A-B) \end{vmatrix}$$

$$= \frac{1}{2n} \begin{vmatrix} 2(A'B')(A+B) & 0 \\ 0 & 2(A'-B')(A-B) \end{vmatrix}$$

The eigenvalues correspond on the one hand to those of the covariance matrix of the sums and on the other hand to those of the covariance matrix of the differences; they are the between- and within-pair eigenvalues. A supplementary dimension is the zygosity. A comparison of the covariance structure of MZ and DZ can be done. The two matrices of the sums can be supposed to be equivalent if MZ and DZ twins are two samples of the same populations (for anthropometric characters, this is not always the case, Defrise, 1982). Estimates of the genetic structure may be found by a comparison of the within-pair covariances.

Another example of the utilization of the techniques for single measurement after the use of multivariate technique is the use of linear vectors (Eaves, 1978; Defrise, 1982). For twins, a matrix transformation is utilized, with

$$M = \frac{1}{n_M}(A-B)'(A-B) \qquad \text{and} \qquad D = \frac{1}{n_D}(S-T)'(S-T)$$

where M pertains to monozygotic twins and D to dizygotic twins. The matrix of the p vectors (p number of characters) formed by the n_M first MZ twins is A, that formed by the n_M second MZ twins is B. The matrix of p vectors formed by the n_D first DZ twins is S, and that formed by the n_D second DZ twins is T. A vector $u' = (u_1, \ldots, u_p)$ is calculated so that the F value of the mean within sum of squares of DZ on the mean within sum of squares of MZ is maximal:

$$F = \frac{u'Du}{u'Mu}$$

is maximal when in

$$(M^{-1}D)u = \lambda u$$

λ is the highest eigenvalue. The λ are compared and tested to define the q eigenvalues which are significantly different. A matrix $U = (u_1, \ldots, u_q)$ of the q corresponding eigenvectors is also defined.

At this level, individual values can be calculated for first and second MZ twins. For a given eigenvector u_j, one computes for MZ twins

$$y_{aj} = Au_j \quad \text{and} \quad y_{bj} = Bu_j$$

and for DZ twins

$$y_{sj} = Su_j \quad \text{and} \quad y_{tj} = Tu_j$$

The classical methodology for single characters can again be applied to these values.

Psychologists have utilized the so-called Multiple Variance Analysis (Cattell, 1960), by comparing different constellations and abstract hypothetical contributions. For instance, for full sibs

$$\sigma^2_{ST} = \sigma^2_{wh} + \sigma^2_{we} + 2r_{wh \cdot we} \sigma_{wh} \sigma_{we}$$

where σ^2_{ST} is the observed variance within sib pairs, σ^2_{wh} the within-family hereditary variance, σ^2_{we} the within-family environmental variance, and $r_{wh \cdot we}$ the correlation between these variances. This kind of approach includes more information than studies of twins, but the method makes oversimplications to reduce the number of unknowns, such as assuming that $r_{wh \cdot we}$ is equal in sibs and dizygotic twins or neglecting factors of interaction. The model makes the assumption of additivity of the genetic and environmental influences and of linearity between inclucences and effects, but it also includes the correlation between the two influences. Cattell et al. (1975) indicate, for instance, a correlation of .25 between genetic and environmental effects on intelligence in terms of between-family variances.

Also working with psychological data, Eaves et al. (1975) developed a model which assumed additive genetic and environmental factors specific for individuals and not for families. The model used for multiple variables is equivalent to the model of Mather et al. (1971) for a single variable. The phenotypic covariance matrix Σ_p is

$$\Sigma_p = \frac{1}{2}(\Delta\Delta' + D^2) + HH' + E^2$$

where Δ and D represent the vector of genetic loadings and the corresponding specific variances, and H and E the vector of environmental loadings and the corresponding variances; D^2 and E^2 are the diagonal matrices. In the study of twins and with matrices of mean products within (W) and between (B) pairs, there are four variables and the vectors Δ and H are four-element column vectors. The following theoretical genetic and environmental contributions are expected:

$$\Sigma_{BMZ} = \Delta\Delta' + D^2 + HH' + E^2$$

$$\Sigma_{WMZ} = HH' + E^2$$

$$\Sigma_{BDZ} = \frac{3}{4}(\Delta\Delta' + D^2) + HH' + E^2$$

$$\Sigma_{WDZ} = \frac{1}{4}(\Delta\Delta' + D^2) + HH' + E^2$$

Eaves et al. (1977) applied this model to the covariation of four aspects of impulsiveness in twins. The data were consistent with the situation of random mating, of additive gene action and environmental effects, and with a single genetic factor with loadings proportional to those of a single common environmental factor.

In a multivariate system, differences between relatives can be described by generalized distances (Defrise, 1967, 1968). The distance, in the normal p-variate population of mean vector p and of covariance matrix Σ, between two points A and B defined by the two vectors $a = (a_1, \ldots, a_p)$ and $b = (b_1, \ldots, b_p)$, is

$$\Delta(A,B) = [(a-b)' \Sigma^{-1} (a-b)]^{1/2}$$

The squared generalized distance between two random points A and B is distributed as $2\chi^2$. Thus, since

$$\frac{\Delta^2(A,B)}{2} \sim \chi_p^2$$

it is possible to test whether the distance between two points is larger or smaller than that between two other points. It is also possible to test whether the whole group of pairs (let us say n MZ, for instance) has more related characters than a group of n random pairs. Under these conditions, when $np \geq 30$,

$$\sqrt{(\Sigma\Delta^2)} = \sqrt{(2\chi^2)} \text{ is distributed as } N(\sqrt{(2np-1)}, 1)$$

The value of the method lies in the fact that not only is a simultaneous study of several measurements possible, but the generalized distances are comparable between subjects of different age and sex. Table 3 gives some results for four head measurements: the difference between $\sqrt{(\Sigma\Delta^2)}$ and the mean of m random pairs $m = \sqrt{(2np-1)}$ demonstrates the similarity of the two relatives. As far as this measure of similarity is concerned, Table 3 shows the intermediate position of the following relatives: DZ, sibs, parent-child. It is evident that identical twins are found much more similar and that the husband-wife similarity is not significant. However, in the intermediate position of the relatives with a coefficient of relationship of 1/2, the similarity seems greater in DZ twins than in sibs, and greater in sibs than in parent-child pairs.

Table 3. Generalized Distances for Head Measurements (p = 4, head height, length, breadth, and frontal breadth)

	Source*	$m = \sqrt{(\Sigma\Delta^2(A,B))}$	$\bar{\Delta}^2(A,B)$	$\bar{P}$	% P<.05
Identical twin pairs (n = 30)	(1)	9.292	1.268	.034	70.0
Nonidentical twin pairs (n = 33)	(1)	6.633	2.784	.127	51.3
Sibs (n = 196)	(2)	5.696	4.604	.215	20.9
Parent-child (n = 552)		6.422	5.209	.251	14.3
Husband-wife (n = 123)		1.919	7.040	.371	6.6

*(1) Defrise, 1970; (2) Susanne, 1972.

It is possible to refine the notion of generalized distance by using a noncentral χ^2. The probability (P) that a random point will lie nearer to A than his relative B can be calculated, A being the point lying the farthest from the center (M); Defrise (1968) showed that when A is given and B chosen at random, the distribution of $\Delta^2(A,B)$ is a noncentral χ^2 with p degrees of freedom and with $\Delta^2(A,M)$ as noncentrality parameter. This methodology has been used for twin studies (Defrise, 1970), for familial study (Susanne, 1972; Susanne et al., 1978), and for assortative mating (Susanne, 1974).

Table 3 gives some results of this method for the same four head measurements which confirm the intermediate position of the relatives DZ, sibs, parent-child, but with a hierarchy of the mean probability $\bar{P}$ DZ < sibs < parent-child.

PATTERNS OF GROWTH

The genetics of growth patterns present additional problems in the collection of suitable data and the quantification of the human growth curve. Consequently there are very few studies on the resemblance of growth patterns.

Most of the studies are qualitative and are done on twins or sibs (for a review, see Susanne, 1980). At a quantitative level, let us mention studies comparing twins or sibs or parent-children at different ages without taking into account the longitudinal aspects present in these studies (see also Susanne, 1980) and studies involving some parameters of longitudinality: Vandenberg et al. (1965) with a linear component of growth rate between 1 month and 4 years and the deceleration during the same period; Welch (1970a,b) with a third-degree orthogonal polynominal between birth and 10.5 years; Fischbein et al. (1978) with the norm of squared differences between corresponding points against the intraclass correlation for profiles estimated from an interaction term of an analysis of variance; Hauspie et al. (1982) with a variance analysis of parameters of the Preece-Baines Model 1 on Bengal sibships.

CONCLUSION

In normally distributed characters, the genotype can only be seen through the phenotype. Genetic and environmental influences seem hopelessly intermingled, but apart from philosophical and political debates, methodological improvements have been made. Some are reviewed here, but it must be evident that the results, whatever the methodology utilized, are relative to the specific environmental conditions of each generation. Heritability or genetic determination has meaning only when the genetic structure and the demographic and the social structure of the population have been defined.

Even the factor time must be taken into account in its influence on heritability, with variation between generations (such as for blood pressure, Ward et al., 1979), variation during growth (such as for height, Tanner et al., 1963), and intraindividual variation (such as for systolic blood pressure, Barcal et al. 1969).

In studies of heritability, we are in a contradictory position: we require a homogeneous population at the genetic and at the environmental level, but also a relatively large sample in order to increase the precision of the values (Lewontin, 1975).

Causes of correlations between heredity and environment are manifold: it can have a genetic origin (genetic factors controling intelligence are causes of upward social mobility; genetic endowment can result in the creation of particular environments more or less protected or ambitious for instance; genetic physical factors such as beauty can provoke specific social treatments), it can have an environmental causation (environmental factors such as more or less protective attitudes can result in differential death rates, environment can stimulate migration), or it can have a common cause (intelligent people sent to better colleges, placement agencies placing adopted children in accordance with the social origin of the child).

Environmental differences are due not only to social differences between families, but geographical factors will contribute as well as income, religion or subculture, schooling, and other more subtle familial factors. Within-family environment is influenced not only by sib position or the different age and experience of parents, but also by considerations outside the family, such as the fact that children attend different schools. The analysis of the total environmental variance remains an essential problem.

It is essential also in path analysis to have a quantification of these environmental contributions. Even if path analysis constitutes a better approach than the Fisher model in the sense that it does not neglect the cultural inheritance, problems are not excluded.

We are in a contradictory situation of recognizing in the models as many factors of variation as possible with the limited number of empirical results, which limits the number of terms for which we can solve, and which moreover reduces the accuracy of the data and of the sampling.

But it is not only to new models that we must turn to get more information on the relative influence of genetic and environmental

effects, but also to new or unusual familial relations. There are
possible familial constellations which are not frequently used, such
as twins or sibs of half-sibs reared apart, unrelated children
reared in the same family or apart, adoptions from near relatives
or unrelated adoptions. There are possible comparisons for cultures
with more or less assortative mating. For random marriages or
marriages of first cousins or incestuous matings, and also comparisons between large and small families, between early and late
marriages, and between sibs with or without age gaps.

Statistically significant results for different traits do not
necessarily indicate independent hereditary control. Multivariate
analysis can treat different variables simultaneously to determine
if they contain independent hereditary components. For instance,
we can get a composite score that gives maximal discrimination between concordant and discordant twins. It is however pure speculation that a géne (as a causal influence) would appear as a factor
in such an analysis or that one factor would represent all genetic
influences and another all environmental influences.

REFERENCES

Barcal, R., Simon, J., and Sova, J., 1969, Blood pressure in twins, Lancet, 1:1321.
Bock, R.D., 1984, Predicting the mature stature of preadolescent children, this volume, page 3.
Bock R.D., and Vandenberg, S.G., 1968, Component of heritable variation in mental tests scores, in: "Progress in Human Behavior Genetics," ed. by S.G. Vandenberg. Johns Hopkins University Press, 233-260.
Cattell, R.B., 1960, The multiple abstracts variance analysis. Equations and solution for nature-nurture research on continuous variables, Psychol. Rev., 67:353-372.
Cattell, R.B., Stice, G.F., and Kristy, N., 1957, A first approximation to nature-nurture ratios for eleven primary personality factors in objective tests, J. Abnorm. Soc. Psychol., 54: 143-159.
Christian, J.C., Ke Won Kang, and Norton, J.A., 1974, Choice of an estimate of genetic variance from twin data, Am. J. Hum. Genet., 26:154-161.
Darlu, P., 1981, La variabilité en physiologie humaine et son interprétation génétique. Application au métabolisme mineral: Mg, Zn, Ca. Doct. Etat. Sc. Natur., Paris VII.
Defrise, E., 1967, Generalized distance in genetic studies, Acta Genetica, 17:275-288.
Defrise, E., 1958, Non-central chi-square in twin studies, Acta Genetica, 18:170-179.
Defrise, E., 1970, Multivariate analysis in twins, Acta Geneticae Medicae et Gemellologiae, 19:150-154.

Defrise, E., 1981, Modèle polygénique pour les caractères mesurables, Bull. Soc. roy. belge Anthrop. Préhist., 92:7-24.
Defrise, E., 1982, personal communication.
Eaves, L.J., 1973, The structure of genotypic and environmental covariation for personality measurements: an analysis of the PEN. Br. J. Soc. Clin. Psychol., 12:275-282.
Eaves, L.J., and Eysenck, H.J., 1975, The nature of extroversion: a genetical analysis, J. Person. Soc. Psychol., 32:102-112.
Eaves, L.J., Martin, N.G., and Eysenck, S.B., 1977, An application of the analysis of covariance structures to the psychogenetical study of impulsiveness, Br. J. Math. Statist. Psychol., 30:185-197.
Falconer, D.S., 1960, "Introduction to Quantitative Genetics." Ronald Press, New York, 365 pp.
Fischbein, S., and Nordqvist, T., 1978, Profile comparisons of physical growth for monozygotic and dizygotic twin pairs, Ann. Hum. Biol., 5:321.
Fischer, R.A., 1918, The correlations between relatives on the supposition of Mendelian inheritance, Trans. roy. Soc. Edinburgh, 52:399-433.
Hauspie, R.C., Das, S.R., Preece, M.A., and Tanner, J.M., 1982, Degree of resemblance of the pattern of growth among sibs in families of West Bengal (India), Ann. Hum. Biol., 9:171-174.
Holzinger, K.J., 1929, The relative effect of nature and nurture on twin differences, J. Educ. Psychol., 20:241.
Iagolnitzer, E., 1977, L'analyse des données appariées. Doct. U.E.R. Math. Univ. Paris V.
Iagolnitzer, E., 1978, Component pair analysis: a multivariate approach to twin data with application to dermatoglyphics, in: "Twin Research: Clinical Analysis." Alan Liss, New York, 211-221.
Jacquard, A., 1977, Concepts en génétique des populations. Masson, Paris, 128 pp.
Lewontin, R.C., 1975, Genetic aspects of intelligence, Ann Rev. Genet., 9:387-405.
Mather, K., and Jinks, J.L., 1971, "Biometrical Genetics: The Study of Continuous Variation." Chapman and Hall, London.
Nichols, R.C., 1965, The national merit twin study, in: "Methods and Goals in Human Behavior Genetics," ed. by S.G. Vandenberg, Academic Press, New York, 231-243.
Rao, D.C., Morton, N.E., and Yee, S., 1978, Resolution of cultural and biological inheritance by path analysis: corrigenda and reply to Goldberger letter, Am. J. Human Genet., 30:445-448.
Susanne, C., 1971, Hérédité des caractères anthropologiques mesurables, Bull. Mém. Soc. Anthrop., Paris, 7:169.
Susanne, C., 1972, Multivariate analysis in genetic studies, Acta Geneticae Medicae et Gemellologiae, 21:204-210.
Susanne, C., 1974, Analyse multivariée dans une étude de l'assortiment matrimonial, Homo, 25:166-171.

Susanne, C., 1975, Genetic and environmental influences on morphological characteristics, Ann. Hum. Biol., 2:279.
Susanne, C., 1980, Developmental genetics of man, in: "Human Physical Growth and Maturation. Methodologies and Factors," ed. by F. Johnston, A. Roche, and C. Susanne, Plenum Press, New York, 221-242.
Susanne, C., and Dash Sharma, P., 1978, Multivariate analysis of head measurements in Punjabi families, Ann. Hum. Biol., 5: 179-183.
Tanner, J.M., and Israelsohn, W.J., 1963, Parent-child correlations for body measurements of children between the ages of one month and seven years, Ann. Hum. Genet., 26:245.
Vandenberg, S.G., 1965, Multivariate analysis of twin differences, in: "Methods and Goals in Human Behavior Genetics," ed. by S.G. Vandenberg, Academic Press, New York, 29-43.
Vandenberg, S.G., and Falkner, F., 1965, Hereditary factors in growth, Hum. Biol., 37:357.
Ward, R.H., Chin, P.G., and Prior, I.A.M., 1979, Genetic epidemiology of blood pressure in a migrating isolate: prospectus, in: "Genetic Analysis of Common Diseases: Applications to Predictive Factors in Coronary Disease," ed. by C.F. Sing and M.H. Skolnick, Alan Liss, New York, 675-709.
Ward, R.H., Chin, P.G., and Prior, I.A.M., 1980, Tokelau Island migrant study: effect of migration on the familial aggregation of blood pressure. Hypertension, 2:43-54.
Weinberg, R., Skear, C.L., Avet, L.M., Frerich, R.R., and Fox, M., 1979, Path analysis of environmental and genetic influences on blood pressure. Amer. J. Epidem., 109:588.
Welch, Q.B., 1970a, Fitting growth and research data. Growth, 34:293.
Welch, Q.B., 1970b, A genetic interpretation of variation in human growth patterns, Behav. Genet., 1:157.
Wright, S., 1934, The method of path coefficients, Ann. Math. Stat., 5:161-215.

EFFECT OF PHYSICAL ACTIVITY

ON FUNCTIONAL GROWTH

Vassilis Klissouras

McGill University, Montreal
Sports Research Institute
Athens, Greece

INTRODUCTION

The traditional approach to the elucidation of the relative influence of physical activity on functional growth is to use either the cross-sectional or the longitudinal study method.

In the first case athletes are compared with non athletes or active individuals are compared with sedentary individuals of the same sex, age and socioeconomic backround. When the longitudinal approach is applied, the same subjects, classified into an experimental and a reference group, are followed up for a period of months or years. In some longitudinal studies a dimensional analysis is added where expected and actual changes in a given biologic attribute as a result of increasing linear dimensions are calculated.

However, these approaches have the obvious limitation that the degree to which the genetic factor is operant in different individuals is not known. For this reason we have used extensively in our studies the twin and co-twin method. From comparisons of intra-pair differences between monozygotic and dizygotic twins, it is possible to separate the relative share of heredity and environment. Monozygotic twins have the same genetic make-up, so that differences in a trait must be due to factors in the environment. Dizygotic twins, on the other hand, have different genotypes and are basically siblings of the same age (Klissouras 1971).

In this symposium I would like to limit my comments on two series of studies we have conducted.

One of them was a co-twin study, where each twin boy was accompanied by his genotypically identical control and this was done in order to obtain some evidence regarding the effect of physical training on functional growth.

The other was a regional study, where we investigated, in their natural environment, boys of a primitive community and city dwellers of different ethnic origin, in order to throw some light on the question: to what extent are variations in functional capacity of adolescents based on genetic differences between populations?

CO-TWIN TRAINING STUDY

Regarding the co-twin training study we should point out that, although we know with some degree of certainty that the trainability of older individuals is smaller than that of those at younger age, we are still largely ignorant of whether increased physical activity is more effective in altering functional growth at an earlier stage of maturation than that at a later one. Thus, we used twelve pairs of identical twins of three developmental ages (10, 13 and 16 years) in order to observe the effects of physical training on maximal oxygen uptake and some polygenic variables related to the oxygen transport system during the growth period (Weber, Kartodiharjo, Klissouras, 1976).

It should be mentioned that the VO_2max is the resultant of a series of cardiorespiratory and metabolic responses during maximal effort and is considered to be the single most important criterion of physical fitness of an individual.

The twin pairs were split (4 at each age) so that one twin trained while his identical brother acted as a control. The training lasted for 10 weeks and was designed to improve primarily the maximal aerobic power. No effort was made to curtail the control twin from his usual kinetic repertoire.

The presentation of the results is limited to some

selected polygenic attributes, which are given in Table 1. The analysis refers to the inter-age and intra-age differences and their statistical significance in the trained and untrained twins before and after training (Table I).

It may be seen that no significant difference was found between trained and untrained 13-year-olds for any of the variables measured. The same holds true for the 10-year-olds, with the exception of their maximal oxygen consumption, which improved more in the trained than in the untrained twins, at a significant level of $P<0.02$. The training had a clear effect on the maximal oxygen uptake, oxygen pulse, ventilation and lactate production in the 16-year-olds.

A statistical comparison of the inter-age difference of differences revealed no significant difference for any of the measurements between 10- and 16-year-olds, and only in maximal oxygen uptake between 10- and 13-year-olds. However, it is obvious that training produced greater changes in several parameters in the 16-year-olds as compared to the 13-year-olds.

The most striking observation was that VO_2max increased by about the same magnitude in both trained and untrained 13-year-old twins. The most likely explanation hinges on the adolescent growth spurt, which occurs at this age as evidenced by the increase in height. The annual increase in body height of the 13-year-olds was 8 cm. A rate of growth of such magnitude is expected to occur only during the adolescent spurt. Whether or not this growth spurt is directly related to growth hormone or to other hormonal factors is still speculative.

It is possible that hormonal activity is maximal during this age and any additional stimuli, such as training, cannot override its influence. In this connection one thinks of the anabolic activity of the growth hormone, which stimulates the transport of animo acids across cell membranes and the synthesis of protein.

However, some other factors must play an essential role, because the blood growth hormone levels in children and adolescents are not different from those observed in adults during rest and in response to muscular work. A raised level of androgen during exercise may act synergetically with the growth hormone.

Some findings from animal studies confirm the hypo-

TABLE 1. Results of selected polygenic attributes of the twins before and after physical training.

POLYGENIC ATTRIBUTE	10-YR-OLDS Δτ	Δυ	13-YR-OLDS Δτ	Δυ	16-YR-OLDS Δτ	Δυ	Δ Δτ - Δυ 10-13	10-16	13-16YR
MAXIMAL OXYGEN UPTAKE ML.KG.MIN	10.61*	4.04	4.16	4.67	8.56*	0.73	0.05	NS	0.02
MAXIMAL OXYGEN PULSE ML.BEAT	2.09	0.66	1.86	2.13	2.84*	0.67	NS	NS	0.02
MAXIMAL HEART RATE BEATS MIN	-1.50	-0.50	-1.00	-0.75	6.25	-2.75	NS	NS	0.05
MAXIMAL WORK VENTILATION LITERS MIN	10.49	8.74	6.56	7.12	26.94*	-10.86	NS	NS	NS
MAXIMAL CARDIAC OUTPUT LITERS MIN	1.03	-0.40	1.83	1.18	1.85	0.20	NS	NS	NS
MAXIMAL BLOOD LACTATE MG + 100 ML	7.25	-0.85	4.13	-14.63	15.00*	-5.95	NS	NS	NS

* CONFIDENCE LEVEL OF 0.05 or more

thesis advanced here that during the growing period, hormones probably play a more dominant role in the development of functional adaptability than does physical activity. Goldberg and Goodman found that growth hormone in normal animal causes proportionately similar increases in skeletal muscles undergoing work-induced hypertrophy and in their controls. When rat muscles (soleus and plactaris) was injected with growth hormone, the muscle appeared to increase in size by the same numeric factor whether there was atrophy from denervation or hypertrophy from increased work. These authors suggest that the hormone determines the absolute changes in muscle size, which result from changes in muscular work.

These comments do not deny that pronounced development observed in adolescents who engage in competitive sports, but simply point out that we have not been able, up to the present, to separate distinctly the relative importance of genetic disposition, growth, and amount of training at different developmental ages.

Several workers have failed to observe a clear improvement in VO_2max in boys who trained during the pubertal growth period.

In an early longitudinal study Parizkova and Sprynarova investigated the somatic and functional development of 96 boys during the growth period (from 11 to 15 years of age). They classified their subjects into four groups according to the intensity of habitual physical activity. For the first two years they observed no difference in the improvement of the aerobic capacity of two groups of boys who engaged in the most and least intensive physical activity regiment. Thereafter, there was a difference, significant at 0.05 level of confidence, when the VO_2max was expressed either in mlmin^{-1} or ml.kg.min^{-1}. When VO_2max was expressed in relative values of LBW such a difference disappeared for the 13-year-olds. Moreover, a closer examination of the data reveals that "the boys who engaged in systematic training already displayed some morphological differences (greater stature and relative heart volume), which may imply a certain disposition for greater activity and performance capacity" (Pariskova 1977). Ulbrich, who studied 130 boys from 11 years up to maturity, reached a similar conclusion (Ulbrich 1971). Ekblom found an absolute increase in VO_2max of 55% for 5 11-year-old boys who had trained for 32 months as compared to 37% for the reference group, but he relative increase was the same (7%) in the two groups. Daniels and Oldridge (1971) also observed in a study of 14 boys aged 11-15 years, who trained for 22 months, that the increase in VO_2max was not at a greater rate than that at which weight increased. At variance with these findings is the report by Eriksson (1972) who noted 16% increase in 12 boys (11-13 years) who trained for 4 months.

It may be argued that the 10-week observation period in our study was too short to relate to the growth spurt the changes noted in the 3-year-olds. For this reason the testing of three pairs of the 13-year-old twins was extended over 1 year. It was found that both trained and untrained twins continued to improve their VO_2max at the same rate until the third testing period. Nevertheless the improvement over this observation period was about 22%, which is substantially more than the improvements noted in the training studies of adolescent boys that have already been cited.

It cannot be decided from the studies mentioned so far whether or not training at prebubescent age can have a greater influence on VO_2max than at a later age. Most probably the ontological time factor is decisive in the development of functionally important structures. However, it remains unscertain to which developmental period the growth promoting stimuli that act upon the tissues should be applied. The old hypothesis that more might be gained by introducing extra exercise at the time when growth impulse is the strongest is no longer tenable in view of the evidence obtained in this study.

A REGIONAL ETHNIC STUDY

A question of considerable interest is whether adaptive responses to maximal effort differ in growing boys of different ethnic origin.

We had the opportunity to study a small number of 51 boys representing four ethnic groups: The Temiars, a primitive community of nomadic hunters inhabiting the jungle, and the Malays, Chinese and Indians who were city dwellers. These three later populations are genetically different in the sense that they are culturally isolated and maintain distinct corporate gene pool through a high rate of endogamy (Klissouras 1978).

We measured the oxygen uptake, blood lactate and heart rate during submaximal and maximal effort. Exercise was done using a step test. In order to control the stepping rate, a metronome was used for the city dwellers and the more familiar bamboo beat for the Temiars. The bamboo beat is used in all-night bomo dances and seems to induce mental concentration. Several dancers were observed to go into trance during maximal testing.

Analysis of the results revealed no significant difference in the parameters tested among the different ethnic groups (Table 2).

TABLE 2. Adaptive responses to maximal effort of adolescents of different ethnic groups

ETHNIC GROUP	OXYGEN UPTAKE $ml.kg.min^{-1}$	HEART RATE beats/min	BLOOD LACTATE mg%
TEMIARS	45.9± 6.7	194± 4	75±18
MALAYS	49.5±10.6	193±10	78±30
INDIANS	47.2± 5.1	198± 5	69±30
CHINESE	43.6± 4.6	196± 3	85±30

It was anticipated that the Temiars might have a high maximal aerobic power as compared to city dwellers. However, although they lead a physically active life, it is at a leisure pace. Telemetric monitoring seldom revealed heart rates exceeding 130 beats per minute during daily activity.

The comparability of functional capacity observed among the different ethnic groups tested is in agreement with several other regional studies where adults were used as subjects. For example, Robinson and his associates (1941) found practically the same maximal aerobic power in American Caucasian and Negro who were living under the same environmental conditions and with similar levels of habitual activity. Wyndham (1966) was impressed with the remarkable similarity of physiological responses between physically active Caucasians, Bantu, and Bushmen in Southern Africa. He noted no differences in VO_2max in spite of the differences between the three ethnic groups with respect to morphology, nutrition and patterns of activity. Davies and co-workers (1972) observed no differences of maximal aerobic power between young adult Kurdish and Yemenite Jews, both groups being active agricultural workers, living under identical conditions, and with similar nutritional state. Neither did they observe any differences between these groups, sedentary Caucasians, and Yoruba villagers.

A mean maximal aerobic power of the same order has been reported for several other ethnic groups including arctic Indians, Eskimo hunters, the Ainu and other Japanese inhabitans of Hokkaido, the Tarahumara Indians of Mexico, Nomadic Lapps, the Pascuans of Easter Island, and Dorobo, Turkana and other African natives (for references, see Klissouras 1978).

It would appear from these various studies that there are no differences in VO_2max among ethnic groups that can be attributed to genetic factors. It must be admitted, however, that the variation within any ethnic group far exceeds the average differences among groups and to compare them in terms of mean values is misleading. Note, for example, that VO_2max expressed in $ml \cdot kg^{-1} \cdot min^{-1}$ ranged from 40 to 62 in the Arctic Indians, from 43 to 84 in the Eskimos, from 29 to 52 in the Ainu, and from 38 to 78 in the Temiars. In fact, one Temiar had a VO_2max of 78, while the respective value for his brother was 47. Both had similar nutrition status and patterns of physical activity, strongly suggesting that genetic factors played an important role in producing the wide difference of VO_2max between them. Our own extensive twin studies, which will be reviewed at the III International Congress of Auxology, in the next few days, strongly suggest that individual differences in VO_2max could be attributed for the most part to genetic differences.

However, what holds true for the individual, does not hold true for the population. Although, more data are needed for a better understanding forman's adaptive variations, it is doubtful whether selective pressures currently prevailing in any biotope are sufficient to cause genetic adaptation, distinct to a population, of the structures and functions which support maximal physical effort.

REFERENCES

Andrew, G.M., M.R.Becklake, J.S.Guleria, and D.V.Bates (1972). Heart and lung functions in swimmers and nonathletes during growth. J. Appl. Physiol. 32:245.

Astrand,P.O., I.Engstroem, B.Erikson, P.Karlberg, I.Nylander, B.Saltin, and C.Thoren (1963). Girl swimmers. Acta Paediat. Suppl.157.

Daniels,J., and N.Oldridge (1971). Changes in Oxygen consumption of young boys during growth and running training. Medicine and Science in Sports, 3:161.

Davies, G.T.M., et al. (1972). Ethnic differences in physical working capacity. J.Appl.Physiol. 33:726.

Ekblom, B. (1969). Effect of Physical Training in adolescent boys. J.J.Appl.Physiol. 27:350.

Erikson, B.O. (1972). Physical Training, Oxygen Supply and Muscle Metabolism in 11-13-yr old boys. Acta Physiol.Scand.Suppl.384

Goldberg, A.L., and H.M. Goodman. (1969). Relationship between growth hormone and muscular work in determining muscle size. J. Physiol., London 200:655.

Greenwood,F.C., W.M.Hunter, and V.Marian (1964). Growth hormone levels in children and adolescents. Brit.Med.J. 1:25-26.

Hunter,W.M., and F.C.Greenwood (1965). Studies on the secretion of human pituitary growth hormone. Brit. Med.J. 1:804.

Klissouras,V. (1971). Heritability of adaptive variation. J.Appl.Physiol. 3:338.

Klissouras,V. Radial differences in functional capacity, in : "Physical fitness assessment". Shephard,R.ed., Charles C.Thomas, Springfield (1978).

Robinson,S., et al. (1941). Adaptation to exercise of negro and white share croppers in comparison with Northern whites. Human Biol., 13:139.

Sprynarova,S.(1966). Development of the relationship between aerobic capacity and the circulatory and respiratory reaction to moderate activity in boys

11-13 years old. Physiol. Bohemoslov. 15:253.

Sutton,J.R. et al. (1973). Androgen responses during physical exercise. Brit. Med. J. 13:520.

Ulbrich,J.(1971). Individual variants of physical fitness in boys from the age of 11 up to maturity and their selection for sports activities. Medicina dello Sport, 24:118.

Weber,G., Kartodiharjo,W., Klissouras,V. (1976). Growth and physical training with reference to heredity, J.Appl. Physiol. 40.

Wyndham,C.H. Southern African ethnic adaptation to temperature and exercise. In: "The Biology of Human adaptability". Baker, P.T. and Weiner,J.S. eds. Oxford, Clarendon Press.(1966).

DISCUSSION REPORT

(Anthropometric Surveys)

>Elizabeth S. Watts
>
>Department of Anthropology
>Tulane University
>New Orleans, Louisiana USA

SUMMARY OF SESSION

   Anthropometric surveys have been, and continue to be, one of our most important sources of data for the assessment of genetic and environmental effects on growth variation in children. While such surveys are usually cross-sectional in nature, and cannot provide direct information on the actual growth process, they can indicate the growth status of groups of children in relation to other variables such as genotype, race, geographical region, socioeconomic class, and nutritional level. Anthropometric surveys may cover very large samples and aim at monitoring child growth in national populations, or they may use more restricted samples chosen to elucidate particular problems. Both kinds of surveys are discussed in the papers in this session.

   The main paper by van Wieringen reports the results of a study of the first type, the large national survey. The extensive growth and maturation surveys undertaken periodically in the Netherlands (in 1955, 1965, and 1980) provide good examples of a kind of study carried out in the post-World War II period in several European countries. (A more complete listing and discussion of surveys of this type can be found in Tanner, 1981, Chapter 14). The Dutch surveys are among the best of these studies, combining careful planning, rigorous standardization and monitoring of measuring techniques, and the inclusion of maturation indicators as well as measures of growth status. These investigations, and others like them, have revealed more clearly the nature and extent of social class differences, and secular changes through time, in size and age of maturity in the children of post-war industrial Europe and North America. The results of the 1980 Dutch survey show that secular increases and social class differences in stature are still evident in late adolescence and presumably in adulthood as well.

Anthropometric surveys have undergone substantial improvements in design, execution, and analysis over the last several decades and have reached a high level of methodological sophistication. Despite the enormous effort and expense involved large-scale national surveys, repeated at appropriate intervals, yield essential information that cannot be obtained by other means. Social class and secular differences in body size and maturation age can result from general trends affecting socioeconomic factors or from specific policies planned and implemented by governments and other agencies. However, it is of first and foremost importance to be able to document whether or not such differences exist, as well as to determine their nature and magnitude. The data can then be used in epidemiological investigations relating differences in growth and development, or the lack of such differences, to specific causes. Anthropometric surveys continue to be a primary means of monitoring the effects of socioeconomic factors on child growth and of assessing their implications for genetic growth potential and the quality of the childhood environment.

The three short papers by Mueller, Susanne, and Klissouras discuss a number of ways in which smaller-scale surveys of anthropometric and other variables can be used to investigate genetic influences on growth and development. These approaches emphasize depth of information rather than breadth and the use of genotypic and genealogical data as well as anthropometry. They illustrate how data on twins, other classes of relatives, and populations, if analyzed imaginatively, can be used to answer basic questions about the nature and degree of genetic influences on child growth and development. Because of our inability to perform breeding experiments and other types of manipulative research with human subjects the survey approach continues to be a major means of investigating the genetics of growth in man.

All three papers show that a variety of models are available and have been used to shed light on the genetics of growth and development when applied to appropriate data from small-scale surveys. However, they also caution that no single model, applied to a given group of children, can yield universally applicable determinations of the genetic components of growth. Each study provides at best only an estimate of the relative influence of genetic factors and all such estimates are population- and situation-specific. In the absence of any universally applicable model capable of assessing genetic contribution the best approach is to look at the results of a variety of models applied to many types of populations and to work towards a consistent picture that can be verified by multiple techniques and data sets.

ISSUES AND RECOMMENDATIONS

The authors of the papers in this session make some specific recommendations regarding the nature and direction of future research using anthropometric surveys and their suggestions will not be repeated here. In the discussion following presentation of the papers, however, a number of questions and issues emerged relating to the design and use of anthropometric surveys in growth research. What follows is a presentation of the major questions raised regarding the assessment of genetic and environmental influences on child growth and some of the directions suggested for future investigations. The workshop participants who contributed to the discussion were, in alphabetical order: B. Bogin, A.R. Frisancho, R. Hauspie, T. Hoang, F.E. Johnston, V. Klissouras, O. Lammert, R.M. Malina, W.H. Mueller, O. Neyzi, A.F. Roche, R.J. Rona, L.M. Schell, C. Susanne, J.M. Tanner, and J.C. van Wieringen.

Major issues regarding large cross-sectional national growth surveys involve their use in assessing the health status of populations and of individuals. While van Wieringen points out that these two purposes provide the primary justification for undertaking such surveys, it is now clear that caution must be used in the interpretation of survey data and in their application to problems of public health and clinical medicine. In the area of public health secular changes and social class differences in average size and maturation age of children are felt to be primarily a reflection of differences in environmental factors. Though it has been suggested that genetic factors are involved in both secular trend (through heterosis) and social class differences (through assortative mating) they have not been shown to play a major role and their effects, if any, are still largely unknown. Thus, the attainment of larger body size and earlier maturation by groups of children is usually attributed to an improved environment, especially in the areas of nutrition and medical care.

Positive secular trend is generally viewed as a positive health indicator for, as van Wieringen points out, it is usually correlated with declining morbidity and mortality during the pre-adult period. However, the meaning of positive and negative secular changes in growth and maturation, or the absence of such changes, can have vastly different health implications depending upon the population involved. For example, continued secular increases in weight, unaccompanied by height increase, in an economically favored nation or social class may simply indicate that the population is becoming fatter, which might confer increased health risks. An absence of secular change could mean that a population has reached its genetic size potential, or it could indicate that environmental conditions are not improving over time (McCullough, 1982). A negative secular trend, or decrease in body size, over time could reflect general decline in socioeconomic conditions (Tobias, 1975). But, slight decreases in some age groups in favored populations may reflect neutral or even beneficial changes as far as overall health status is concerned. In the 1980 Dutch survey van Wieringen feels that the slightly smaller infant size may possibly be related to restrictions on maternal weight gain during pregnancy and changes in infant feeding practices, such as increased breast feeding and reduction in the fat and sugar content of infant formulas. Interpretation of secular change, therefore, should always take into account the type of population and the conditions under which it is currently living and has lived in the past.

In using secular and social class differences in growth status as measures of health status and environmental adequacy an important question is how to determine when a given population has reached an optimum body size with respect to its environment. Mueller suggests that one way to approach this question is to determine morbidity and mortality of children in each growth percentile for the population in question. One could then determine a point at which increased size carries increased health risks in that population.

From a global perspective perhaps the most far-reaching aspect of this issue is the use of child growth and body size as a measure of the quality of life experienced by populations and segments of populations throughout the world. Are the maximization of secular trend and the elimination of environmentally caused differences in body size a desirable goal? To physicians and public health workers any degree of growth retardation is viewed as bad, while to the population biologist it may be seen as a favorable adjustment to existing environmental conditions. Stini (1975) has pointed out that larger children growing into larger adults require more calories and the maximization of growth potential may be counterproductive for the population if it cannot procure the calories needed to support the resulting larger biomass. Here again it is clear that secular trend must be considered in relation to demographic features of populations, as well as other aspects of their health and nutritional environments, before its significance can be fully evaluated.

A second issue concerning data from large cross-sectional surveys is their use in the construction of reference standards for evaluating the growth progress of individual children. Tanner (1981) and others have repeatedly warned against the pitfalls of using purely cross-sectional data for this purpose, the primary reason being that the cross-sectional means tend to underestimate growth when it is most rapid. Longitudinal data, analyzed longitudinally, with individual tempos taken into account, are far superior for this purpose. However, since such data are not available in large quantities nor for all human groups a compromise is necessary. This involves using longitudinal data to make appropriate adjustments in the cross-sectionally derived curves thereby maximizing their accuracy as standards (Tanner et al., 1966, Tanner and Whitehouse, 1976). In gathering cross-sectional data for this purpose it is also useful to increase sample size at ages when growth is most rapid (Jordan et al., 1975) and include measures of maturation status as well as size.

It is now evident that surveys designed to elucidate the genetics of growth must go beyond simple assessments of heritability of attained size. In fact, the papers in this session show that many researchers have already begun to deal with more complex types of relationships taking into account various kinds of genetic influences operating on a variety of developmental variables such as physiologic function and performance as well as size.

Two points were raised concerning the methodology of growth genetics research. The first is the fact that none of the methods currently available can truly separate genetic from environmental influences on growth. At both the population and family levels there are interactions between geneotype and environment that confound attempts to separate the two. In many instances the closer the genetic relationship between people the more similar their environments. Susanne points out that this principle extends right down to the level of monozygotic versus dizygotic twins. Assortative mating for characters like stature can also confound heritability estimates. Thus, the fact that both the genetic structure and the degree of environmental heterogeneity may vary from group to group means that assessments of the genetic component of growth cannot be absolutely determined at present.

A second methodological point expressed is the need for surveys to define more clearly those features of the environment relevant to growth and to devise more accurate measures of these variables. It is clear that the same set of variables will not be relevant in all populations. Even though there may be common factors affecting the growth process in most populations they may have to be measured in different ways.

Several discussants emphasized the need to take a closer look at the growth process in terms of stages rather than as a whole, and at the growth responses of individuals rather than the mean group response to particular environmental influences. Johnston raised the question of whether growth may be under stronger genetic control during the adolescent period than during preadolescence. While this issue is far from being settled it does point up the necessity of considering the genes-environment relationship separately for each phase of growth. Mueller and Frisancho both cited evidence from very different areas of research suggesting that individuals in a single population may make different growth responses to a given environmental factor. Such differential responses, which may be gene-related, would be missed when only the mean group response is considered.

An understanding of growth phase differences and individual differences in environmental sensitivity would greatly increase our understanding of genetic and environmental interaction in the growth process as a whole. Better knowledge of these differences would also have important benefits of a more practical nature as well. A given environmental insult may produce growth retardation in some individuals but not others, and at some phases of growth. Conversely, remedial measures, such as nutritional supplementation, designed to alleviate growth retardation may benefit individuals in different ways and may be more beneficial during particular phases of growth. Clearly, such possibilities call for more carefully designed surveys and analyses that address these questions specifically.

REFERENCES

Jordan, J.R., Ruben, M., Hernandez, J., Bekelagua, A., Tanner, J.M. and Goldstein, H., 1975, The 1972 Cuban national child growth study as an example of population health monitoring: design and methods, Annals of Human Biology, 2:153.

McCullough, J.M., 1982, Secular trend for stature in adult male Yucatec Maya to 1968, Am. J. Phys. Anthropol., 58:221.

Stini, W.A., 1975, Adaptive strategies of human populations under nutritional stress, in: "Biosocial Interrelations in Population Adaptation," E.S. Watts, F.E. Johnston and G.W. Lasker (eds.), Aldine, Chicago, p. 19.

Tanner, J.M., 1981, "A History of The Study of Human Growth," Cambridge University Press, New York.

Tanner, J.M. and Whitehouse, R.H., 1976, Clinical longitudinal standards for height, weight, height velocity and weight velocity and the stages of puberty, Archives of Disease in Childhood, 51:170.

Tanner, J.M., Whitehouse, R.H., and Takaishi, M., 1966, Standards from birth to maturity for height, weight, height velocity, and weight velocity: British children, 1965, Archives of Disease in Childhood, 41:454.

Tobias, P.V., 1975, Anthropometry among disadvantaged peoples: studies in Southern Africa, in: "Biosocial Interrelations in Population Adaptation," E.S. Watts, F.E. Johnston and G.W. Lasker (eds.), Aldine, Chicago, p. 287.

Medical Environment

INTRODUCTION

Machteld Roede

University Children's Hospital
"Het Wilhelmina Kinderziekenhuis"
Utrecht, The Netherlands

How may we define "medical environment"? It does not necessarily imply the study of children who have lived for a long time in an institution because of some chronic illness, or of those who need extensive medical treatment. Here, we want to show how knowledge about the patterns of normal growth and its variation could be related to both clinical work and public health care. Of particular importance is the practical application of actual national growth standards in the general screening of children by health services teams.

It has been well established that the socioeconomic and ecological environment can have a lasting influence on growth and development: nonoptimal conditions result in inhibiting effects. The term "medical environment" can be considered as covering the effects of medical care given to both the mother and the child, not only in cases of disease, but also the health care given to all children in a population. The discussion below comprises examples how inhibiting factors can be either removed, diminished, or prevented.

Growth is an extremely sensitive parameter of the well-being of a child. Therefore regular measurements are to be encouraged. This section includes some general statements about advisable applications in population health care and about handling of groups at risk. A normal screening routine should at least include measurements of supine length/height and weight, and for individuals who are older than eight years ratings of sexual development. For infants, it is also adivsable to measure the head circumference at regular, short intervals. Since a great proportion of head growth takes place during the first year of life, information on head circumference offers a most sensitive indicator of a child's general growth process. Each (lasting) deviation from the normal growth channel - in either supine length/height, weight, or head circumference - should alert the physician to look for possible causes of retardation. It has been shown that deviations of the growth channel often become apparent before clear symptoms of the underlying illness and/or environmental stress are obvious.

It should be emphasized that supervision of growth should start not at birth, but during the antenatal period - during pregnancy. It will be discussed below how important the follow-up of pregnant women is, e.g., for early diagnosis of fetal growth retardation. Appropriate medical care can help children to achieve their optimal growth pattern, even when their prenatal conditions have not been optimal.

Also, the importance of immediate medical assistance to growth-retarded children in order to offer them a chance to catch up will be discussed, as well as the effects of malnutrition.

The pre- and postnatal period and the first two to three years of life are a most vulnerable period of development, because of the high growth velocity, rendering the child most sensitive to influences of the environment. Therefore, we shall concentrate in our discussion of the subject on early childhood. But it should be realized that during puberty, another developmental period of high growth velocity, when so many changes in the body take place, the inhibiting effects of the environment on the growth process might be significant. Consequently, medical care should by preference be prolonged until after this period.

The co-workers in this session sincerely hope this discussion on "medical environment" will reach not only auxologists but also many other clinicians, and that it may stimulate the inclusion of more and better knowledge on growth and development in medical training.

HUMAN GROWTH UP TO SIX YEARS - DISTANCE AND VELOCITY

Ingeborg Brandt

University Children's Hospital Bonn
5300 Bonn 1
Germany (FRG)

INTRAUTERINE UP TO 40 POSTMENSTRUAL WEEKS

It is important to consider intrauterine growth for various reasons: Firstly, such standards are a prerequisite for understanding the patterns of intrauterine growth. They are indispensable for judging the timing and extent of fetal growth retardation and of postnatal catch-up growth. Secondly, growth standards of the fetus are necessary in order to compare the postnatal extrauterine growth of preterm infants until their expected date of delivery.

Normal growth can be considered a parameter of well-being of the fetus. Today, the timing of growth retardation of the fetus can be well documented by follow-up ultrasound examinations during pregnancy which have produced a real "break through" (Hansmann 1974, 1976).

The emphasis of my report will be on growth of the fetus in the last trimester of pregnancy, i.e., from 28 postmenstrual weeks to term. During this period the intrauterine growth may be affected by malnutrition from different causes, and the rapidly growing fetus does not behave as a parasite, as it is often the prevailing opinion.

The main factors contributing to fetal malnutrition may be seen in table 1 which gives an etiological classification of intrauterine growth retardation.

It should be noticed that placental factors (b) are closely interwoven with the maternal factors (c) in Table 1. Intrauterine growth retardation may occur in infants of mothers with gestosis and hypertension or of smoking mothers for example.

Table 1. Etiological classification of intrauterine growth retardation

a) <u>Fetal factors</u>
   genetic defects
   chromosomal abnormalities
   malformations
   infections
   radiations
   hypoxemia

b) <u>Placental factors</u>
   extremely small placenta
   unfavourable site of implantation
   anomalies of form and of cord
   disturbances of differentiation
   severe maternal disease affecting the placenta

c) <u>Maternal factors</u>
   "vascular insufficiency" mechanisms
   gestosis
   multiple pregnancy
   genetically small mother
   smoking
   undernutrition
   low social class

The prognosis of SGA infants caused by factors from group (b) or (c) is dependent not only of duration, timing, and severity of the intrauterine restriction but also of immediate postnatal care and especially of nutrition.

The significance of the problem of intrauterine growth retardation results from the incidence of about one third SGA infants among all low-birth-weight infants in industrialized societies (Scott and Usher 1966; Brandt 1978).

<u>Definitions.</u> (a) Postmenstrual age, synonyms are fetal or gestational age, is the time from the first day of the mother's last menstrual period until birth. (b) Infants with normal intrauterine development, i.e., with a birthweight between the 10th and 90th percentiles of Lubchenco et al. (1963) or national standards are defined as appropriate-for-gestational age (AGA); small-for-gestational age (SGA) infants are defined as those with a birthweight below the 10th percentile (Brandt 1978). (c) Preterm infants are defined as those born with a postmenstrual age of less than 38 weeks irrespective of birthweight (McKeown and Gibson 1951; Battaglia and Lubchenco 1967). The other definition of preterm infants, as those with a postmenstrual age of "less than 259 days or 37 completed weeks" (Dunn 1979) is not recommended since infants who are born 21 days before regular term (280 days) cannot be considered, and treated, as "mature"

(Brandt 1978). It is important to remember that 40 weeks gestation, or 280-286 days, is the 41st week (NOT the 40th) since the first day of the last menstrual period. (d) Low-birth-weight infants (LBW) means a birthweight of less than 2500 g (Dunn 1979), irrespective of duration of gestation. This very simple definition is of value in identifying most high risk infants (AGA preterm as well as SGA preterm or full term) when information on gestational age is missing.

General Growth Patterns

Liberalization of abortion laws has offered an opportunity for the study of the correlations between postmenstrual age and the rate of early bodily development of human embryos and fetuses.

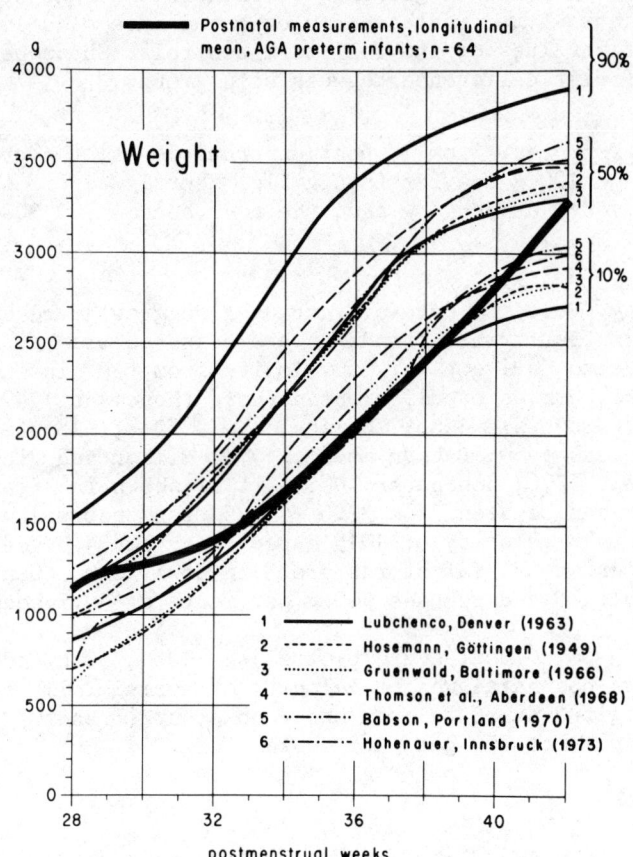

Fig. 1. Mean weight curve of the Bonn AGA preterm infants, postnatal measurements, longitudinal, compared with "intrauterine" standards from different authors (Brandt 1978).

Weight

Distance. The data analysis of a co-operative study in the USA, Denmark and Hungary, based on 783 specimens between 6 and 20 weeks' gestation indicated that in the first twenty weeks of pregnancy the rate of fetal weight increase is substantially slower than it had been assumed previously by the anatomists (Iffy et al. 1975).

Postnatal growth of AGA preterm infants until their expected date of delivery: Figure 1 shows the mean weight curve of the AGA preterm infants, plotted against the intrauterine standards of different authors. The extrauterine weight gain is smaller than that suggested for the fetus in utero. In the corresponding period from 33-40 postmenstrual weeks the weight curve of the AGA preterm infants follows approximately the 10th percentile of the "intrauterine" standards (Brandt 1978). This growth retardation is caught-up by the AGA preterm infants at two months after expected date of delivery, when they reach the values of the full term controls. Subsequently there is no significant difference between both groups.

Sex differences. From 30 postmenstrual weeks, the preterm boys are on average 100 g heavier than girls. Shortly before the expected date of delivery, at 38-40 weeks, the boys become 200-300 g heavier ($p < 0.025$).

Velocity. Velocity data of intrauterine growth are not available since "intrauterine" standards are cross-sectional. An attempt has been made to derive a weight velocity from the distance data of "intrauterine" curves of different authors (Hosemann 1949; Lubchenco et al. 1963; Gruenwald 1966; Thomson et al. 1968; Kloosterman 1969; Usher and McLean 1969; Babson et al. 1970; Milner and Richards 1974; Brenner et al. 1976; Hohenauer 1980), it is shown in Figure 2. A growth spurt can be seen from 32.5 - 36.5 postmenstrual weeks with a maximum weight velocity of 1015 g per month at 34.5 weeks. During this spurt period the fat stores are laid down. Subsequently the velocity declines and reaches 500 g per month at 39.5 weeks.

From 0.5 to 36 months are plotted the 10th, 50th and 90th percentiles of the postnatal weight velocity of normal full term infants, based on longitudinal data from the Bonn study (Brandt 1978). There is a median peak of 950 g at 1.5 months (Fig. 2).

Supine Length

Distance. In the earliest stages of development the sitting height or crown-rump length (CRL) is taken as a growth parameter. It possesses additional advantages in that in fetuses it can be determined more accurately than the crown-heel length.

MEDICAL ENVIRONMENT

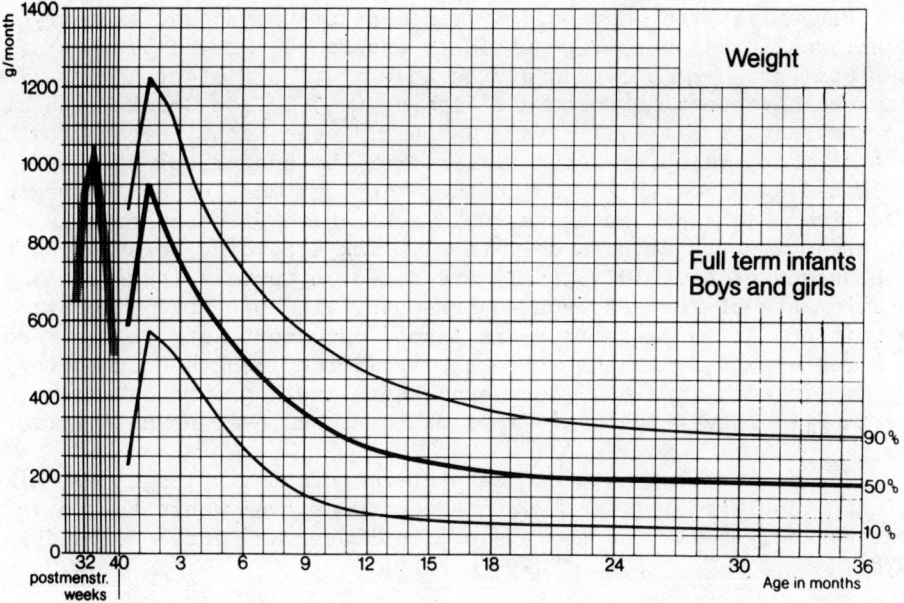

Fig. 2. Velocity curve of weight, calculated in g/month; from 30.5 - 39.5 postmenstrual weeks the 50th percentile, derived from "intrauterine" curves; from 0.5 - 36 months the 10th, 50th and 90th percentile, based on longitudinal measurements of full term infants (Brandt 1978).

Table 2. Growth in the First Weeks of Pregnancy

Postmenstrual age in weeks	Crown-rump length in mm		
	Streeter (1920)	Robinson et al. (1975)	Hansmann et al. (1979)
8	23	14.7	10.8
9	31	21.9	15.4
10	40	30.5	21.5
11	50	40.4	32.4
12	61	51.7	43.2
13	74	64.3	55.6
14	87	78.3	86.3
15	101	—	101.1
16	116	—	105.0

Recently, with increasing refinement of ultrasound biometry techniques it has become possible to assess the growth pattern of the fetal crown-rump length during the first half of pregnancy. The pioneers have been Robinson and Fleming (1975) from Great Britain, and Hansmann et al. (1979) from Germany. These actual sonar growth data of the developing fetus differ from the anatomical values published by Streeter (1920) in that they are smaller at the beginning (Tab. 2). At 10 postmenstrual weeks where Streeter reports a CRL of 40 mms, it amounts only to 21.5 mms according to actual data of Hansmann et al. (1979); they are based on (partly longitudinal) measurements of fetuses from sonographs giving pictures of better definition. At 15 weeks there is a good agreement between the results of Streeter and Hansmann, where the crown-rump length amounts to 101 mms in both studies. That means that length growth in early pregnancy - like weight - is less rapid as has been assumed formerly.

The postnatal growth of AGA preterm infants in length is, like weight clearly small; from 31 to 40 postmenstrual weeks the curve runs markedly below the average of "intrauterine" curves (Brandt 1978).

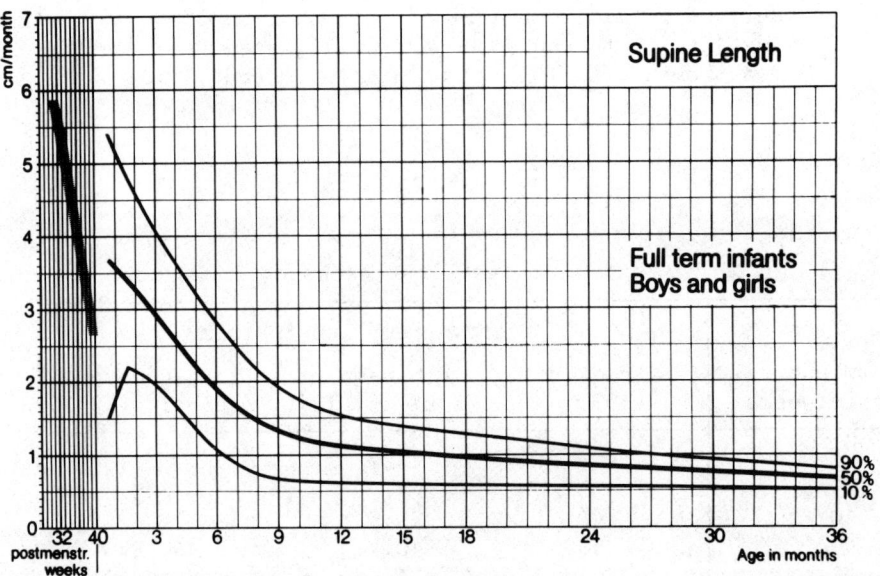

Fig. 3. Velocity curve of supine length, calculated in cm/month; from 30.5 - 39.5 postmenstrual weeks the 50th percentile, derived from "intrauterine" curves. From 0.5 - 36 months the 10th, 50th and 90th percentile, based on longitudinal measurements of full term infants.

<u>Velocity</u>. During the early part of pregnancy from 11 to 18 postmenstrual weeks there is a period of very high growth velocity of crown-rump length; it amounts to 10 - 12 mm per week with a peak of 12.6 mm in the 14th postmenstrual week (Hansmann et al. 1979). Throughout the remainder of the fetal period growth in length is surprisingly constant; the fetus grows with a constantly decreasing rate.

For the third trimester of pregnancy length velocity has been derived from the distance data of "intrauterine" curves (cross-sectional), as published by different authors (as in Fig. 2 above). Figure 3 demonstrates the 50th percentile of length velocity, calculated in cm per month; a steady decrease can be seen from 5.8 cm at 30.5 weeks to 4.1 cm at 36 weeks and to 2.0 cm at 39.5 weeks.

Between 0.5 and 36 months are plotted in figure 3 the 10th, 50th and 90th percentiles of length velocity of normal full term infants. There is a median peak of 3.7 cm at 0.5 months, subsequently the curve flattens; at six months velocity decreases by 50 per cent to 1.8 cm (Brandt 1978).

Head Circumference

Head circumference is an important growth parameter. Its close relationship to brain development in the first two years has been shown in many studies (Brandt 1981b). Repeated postnatal measurements of head circumference - or intrauterine assessment of the biparietal diameter - are of prognostic value; they can reveal a slowing of skull growth and therewith a retardation of brain development.

<u>Distance</u>. The postnatal growth of head circumference of AGA preterm infants until their expected date of delivery is similar to that of the intrauterine fetus (Brandt 1978). After term the head circumference growth of AGA preterm infants is identical with that of the full term controls. Contrary to weight and length, growth in head circumference of the preterm infants is not diminished.

There are already significant sex differences in head circumference of 0.6 to 1.4 cm with boys larger from 30 postmenstrual weeks. After term the boys' head circumference continues to be 0.8 - 1.5 cm larger, a significant difference ($p < 0.01$).

<u>Velocity</u>. Velocity curves of intrauterine head circumference growth do not exist because of the technical difficulties already described. By regular weekly measurements (longitudinal) of AGA preterm infants who are born at least two months before expected date of delivery the growth velocity in the third trimester of pregnancy could be obtained (Fig. 4).

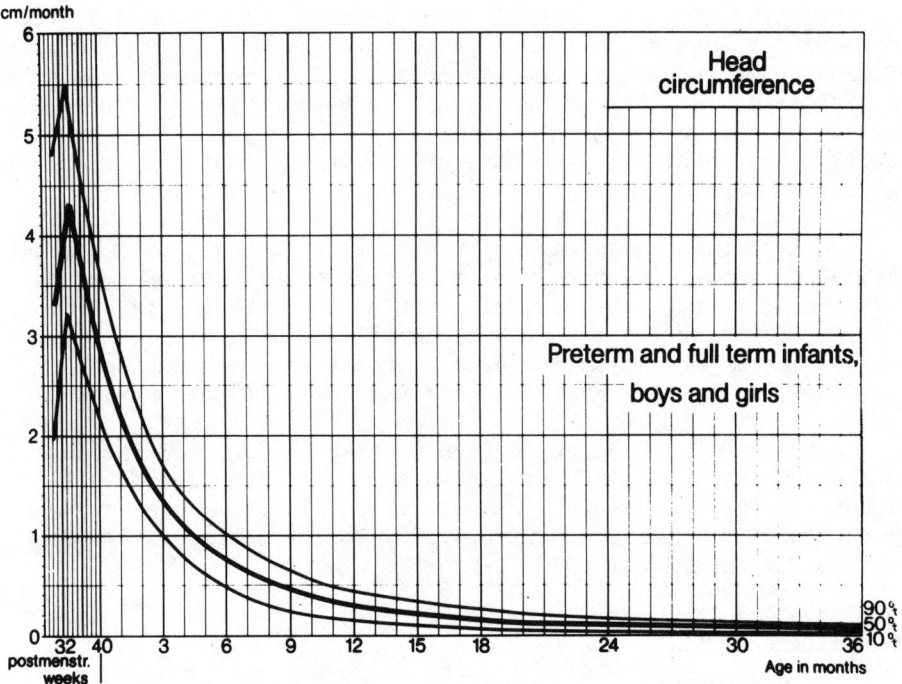

Fig. 4. Velocity curve of head circumference, calculated in cm per month, from 30.5 postmenstrual weeks to 36 months after term, the 10th, 50th and 90th percentiles based on longitudinal measurements of AGA preterm and full term infants.

From 30.5 to 39.5 postmenstrual weeks the growth velocity of head circumference, calculated in cm/month, is significantly higher than after term (Fig. 4). It amounts to 3.3 cm in the 31st postmenstrual week, reaches a peak of 4.3 cm in the 34th week and decreases to 3 cm in the 40th week (Brandt 1978). After term the velocity decreases rapidly, at 18 months it amounts to 0.2 cm/months, these are only five per cent of its peak velocity.

Conclusions. Intrauterine growth standards are indispensable for the assessment of intrauterine development of the fetus according to his postmenstrual age, and for differentiation of newborn infants whether they are appropriate, large or small for their gestational age. Such a classification is always necessary because of the different risks of these three groups in the immediate neonatal period, and for planning the postnatal care and nutrition.

POSTNATAL GROWTH FROM BIRTH TO SIX YEARS

First, let us consider normal full term infants compared with appropriate-for-gestational age (AGA) preterm infants. Growth in the first two to three years is of special interest because of its high velocity and because of its greatly different patterns from child to child. At birth the growth dynamics of a normal child are difficult to predict; it is not possible before the age of 3, when the full correlations with his parents' height or his own adult height are established, with a correlation coefficient of 0.81 for the boys and of 0.75 for the girls (Tanner and Whitehouse 1982).

Besides the hereditary factors, final height is influenced by the nutritional and other environmental conditions during the first year of life, a period of very rapid growth. This is demonstrated in figures 2, 3 and 4 above where the growth velocity of weight, length and head circumference is plotted from birth to 3 years. I should like to suggest that the phenomenon of secular acceleration is mainly based on the environmental conditions during the first one or two years.

Methods

The sample. The growth patterns presented here are based on a longitudinal study from birth to the sixth year of age. The final sample consists of (a) 84 full term infants chosen at random; and (b) 115 preterm infants with birthweights of $\leq$ 1500 g, admitted to the Bonn University Children's Hospital between 1967 and 1975, - because of their small number all SGA infants with birthweights $\leq$ 1500 g were still enrolled until 1978 -, and some of higher birthweight, chosen at random. In Bonn, the small preterm infants of the surrounding obstetric departments are cared for at the University Children's Hospital. Infants with malformations and with severe neurologic handicaps have been excluded.

Using the criteria of postmenstrual age and birthweight, the preterm infants are classified into 64 AGA infants and 51 SGA infants. Each of the latter had a birthweight below the 10th percentile of Lubchenco et al. (1963).

For calculation of the growth curves of preterm infants the age has to be corrected; that is, discounting the time of prematurity from chronological age. To achieve such an age correction the assessment of postmenstrual age has been particularly emphasized in our study (Brandt 1978).

General Growth Patterns

Weight

Distance. The birthweight (50th percentile) of the full term infants of the Bonn study amounts to 3470 g for the boys and to 3300 g for the girls. This is in accordance with other studies (Brandt 1978). Birthweight has shown no secular trend.

At 40 postmenstrual weeks the weight of the AGA preterm infants is significantly less than that of the full term controls by 500 g, and at one month by 300 g. From two months there is no significant difference between both groups. Therefore the weight percentiles in figure 5 from term to 2 months are based on full term infants alone (n=84) and from 3 months to 5 years on AGA preterm and full term infants combined (n=148) (Brandt 1979b, 1980b).

International comparisons including other European studies and the USA study (National Centre for Health Statistical) show similar results to those of the Bonn infants with a tendency to be less heavy.

Velocity. The weight velocity of the fullterm infants, calculated in g per month, amounts to 580 g (50th percentile) in the first month, reaches a peak of 945 g (50th percentile) in the second (Fig. 2 above), decreases to 545 g in the 6th month, and to 200 g in the 21st. Subsequently the weight velocity remains relatively constant at about 180 g per month, i.e., 2.2 kg per year (50th percentile), similar to the Swedish study of Karlberg et al. (1976). Knowledge of the postnatal weight velocity is also of great practical significance. Young and less experienced mothers are often troubled by a weight gain of 1000 g per month by their baby during the first and/or second month. This quite normal weight gain corresponding to about the 50th percentile often induces the mother to reduce the amount of food; this happens even if the baby is breast fed. The baby in his turn is troubled and cries in protest against this food restriction, once more troubling the mother and the rest of the family.

Supine Length/Height

Distance. The length at birth (50th percentile) of the full term infants amounts to 52 cm for boys and to 51 cm for girls. Also in the following months up to six years there remains a sex difference in supine length/height of 0.5 to 2 cm in favor of the boys (Fig. 6).

At 40 postmenstrual weeks (term) the length of the AGA preterm infants is significantly less than that of the full term controls by 2.8 cm; here the 50th percentile of the preterms corresponds to

MEDICAL ENVIRONMENT

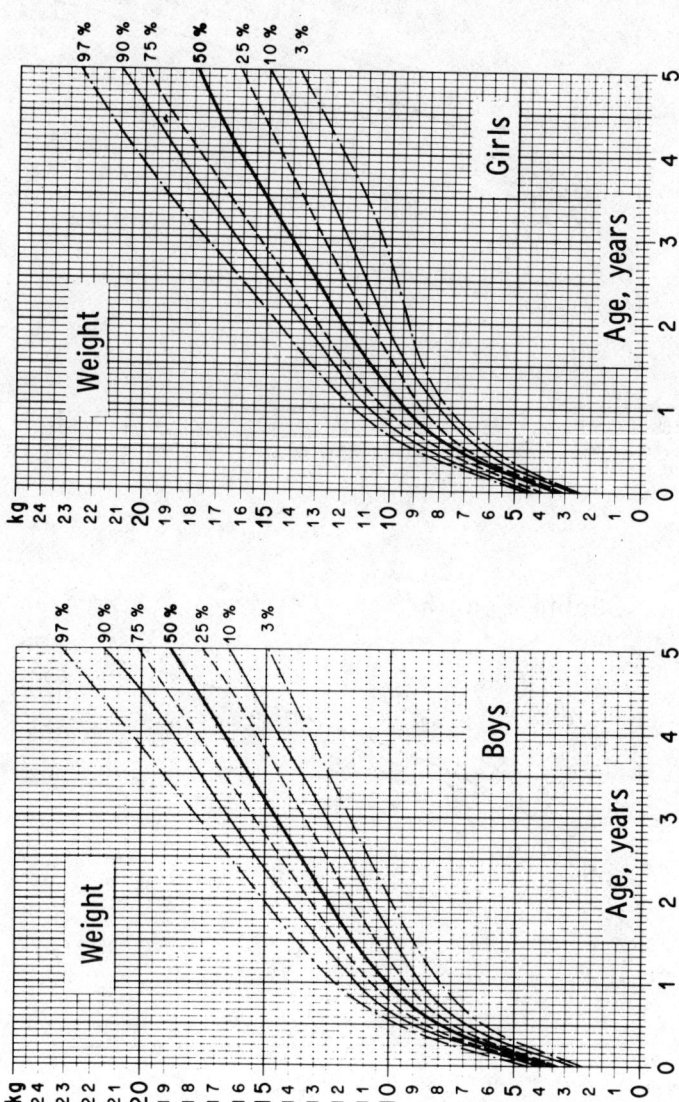

Fig. 5. Weight percentiles, from term to 2 months full term infants alone, from 3 months to 5 years AGA preterm and full term infants combined, boys left, girls right.

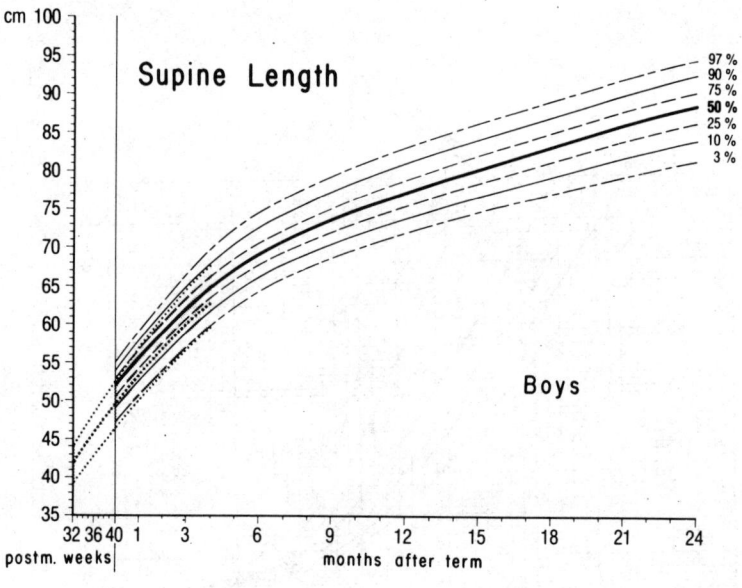

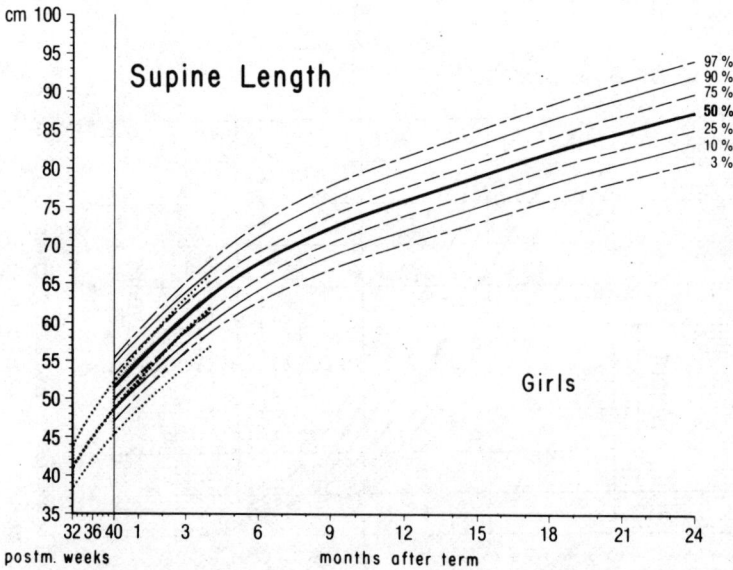

Fig. 6. Supine length: From 32 postmenstrual weeks to 4 months the 3rd, 50th and 90th percentiles of the AGA preterm infants (dotted lines), from term to 24 months the 3rd to 97th percentiles of full-term infants.

about the 10th percentile of the full terms (Fig. 6). The difference decreases to 1.2 cm at 18 months; from 21 - 60 months it amounts to 0.3 - 1.2 cm and is no longer significant.

The standards for supine length are based on the full term infants alone from 0 to 24 months (Brandt 1978, 1980a) (Fig. 6).

International comparisons show that the Bonn children are taller than the children of most other European studies and of similar stature to Dutch infants (van Wieringen et al. 1971) who are among the tallest in the world (Brandt 1978, 1980b). The median of the Bonn study corresponds well with the 75th percentile of the growth charts of Tanner et al. (1966). A similar trend continues to exist up to 16 years for another German study (mixed longitudinal) undertaken in Dortmund from 1968 to 1978 (709 boys, 711 girls aged 1.5 to 16 years) (Reinken et al. 1980). Recently these data have been coupled with those of the Bonn study.

A comparison with the Hagen study of the "Nachkriegskinder", born in 1945/46 (Hagen 1966) from six to sixteen years shows that the 50th percentile of the Bonn/Dortmund studies follows the 75th of the Hagen study. The height difference between both studies, i.e. the secular trend, ranges between 4 and 6 cm for boys as well as for girls.

Velocity. The length velocity of the full terms, calculated in cm per month, amounts to 3.7 cm in the first month (Fig. 3 above), decreases to 1.7 cm at 6.5 months, to 0.9 cm at 22.5 months, and to 0.6 cm per month at 3.5 years (Brandt 1978); length velocity has its peak earlier in fetal life.

Head Circumference

Distance. Today the close correlation between head circumference growth and brain development in the first two years of life is no longer disputed (Dobbing and Sands 1978).

To determine if a child's head size is too big, too small or normal, the head circumferences of the father and mother are important. Actual publications on heritability (father/son, mother/daughter correlations etc.) of head circumference do not appear to exist (Brandt 1981b, Susanne 1982). According to the results of Weaver and Christian (1980) one half or more of head size variation seems to be familial.

For differential diagnosis hydrocephalus/familial megalencephaly the hereditary background of the child is of great importance. In the study of Lorber and Priestley (1981) of 109 children with megalencephaly there was a familial incidence of at least 50%.

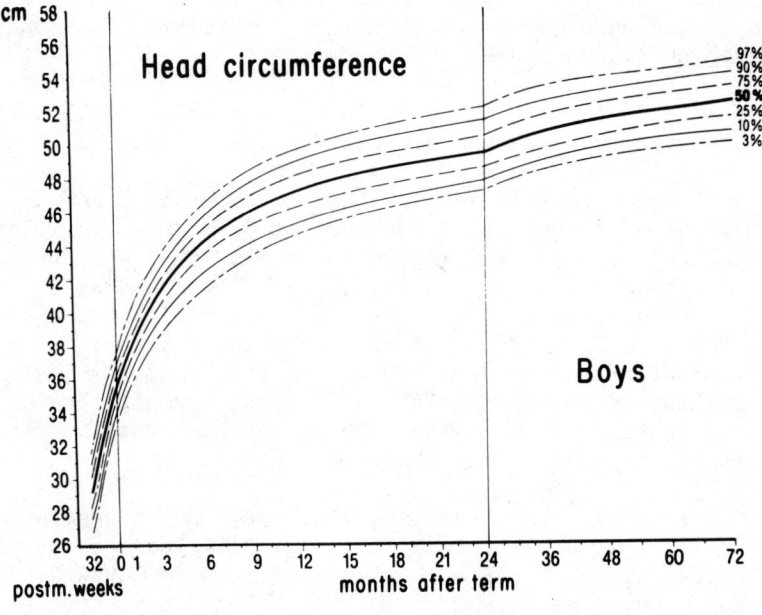
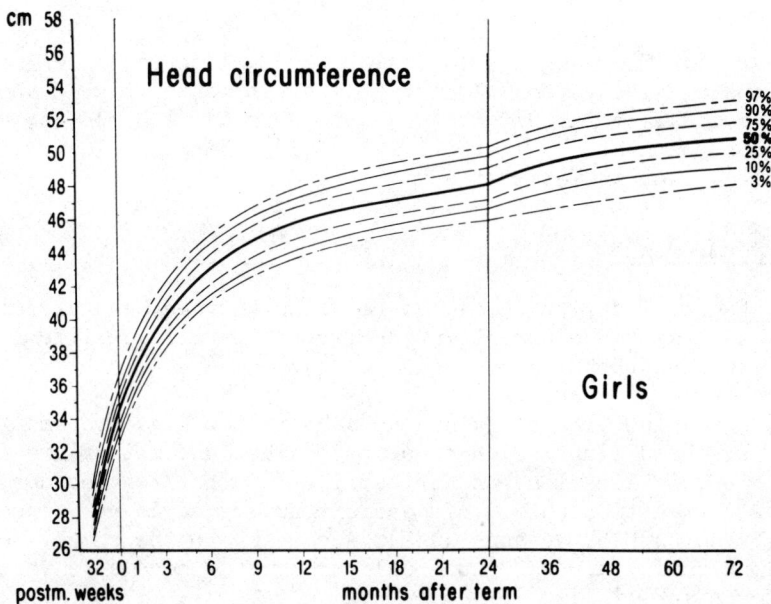

Fig. 7. Head circumference: From 32 - 39 postmenstrual weeks AGA preterm infants, and from 40 postmenstrual weeks (term) to 72 months AGA preterm infants and full term infants combined, boys above, girls below.

In comparison with height the secular acceleration of head circumference is only small (Meredith 1946). There exists a trend to a greater head circumference but the increase of 3-4 mm is less than one tenth than that of height.

From term to six years the head circumference of the AGA preterm infants agrees with that of the full term controls (Brandt 1978). In both groups also the mean head circumference of the fathers (of preterms 57.3 cm, SD 1.6, and of full terms 57.6 cm, SD 1.7) and of the mothers (of preterms 54.7 cm, SD 1.3, and of full terms 54.8 cm, SD 1.5) agrees well. The standards for head circumference in figure 7 are based on AGA preterm infants and full term infants combined (n=148) from 40 postmenstrual weeks to 72 months (Brandt 1979a, 1980b).

International comparisons show a close agreement of the Bonn results with those of children of other European studies and of the USA (Brandt 1978, 1980b).

Sex differences. At term the mean head circumference is 35.4 cm in boys and 34.6 cm in girls, a significant difference of 0.8 cm ($p < 0.01$). Up to the age of six months the difference increases to 1.4 cm and remains at about this level until adulthood (Brandt 1981).

Dolichocephaly. The question of the influence of head shape on head circumference in preterm infants has given rise to different attempts at correction (Baum and Searls 1971a+b; Largo and Duc 1977). Using the correction formula of Baum for example did not essentially change the Bonn results, because by this correction the head circumference of the full term infants decreased to nearly the same extent as it did that of the preterms. From this it follows that the proportion of the compared groups remained unchanged (Brandt 1980b).

A comparison of the areas surrounded by the head circumference of preterm and full term infants, calculated from biparietal and fronto-occipital diameters of the infants with the elliptical equation yields the following results (Tab.3): At 40 postmenstrual weeks the area surrounded by the head circumference agrees surprisingly well in preterm and full term infants (84.7 versus 84.5 square cm).

Table 3. Skull Measurements at 40 Postmenstrual Weeks (Term) in cm, Boys and Girls

	Fronto-occip. diameter (AP)		Biparietal diameter (BP)		Head circumference		Area*
	M	SD	M	SD	M	SD	
Full term infants	11.7	0.5	9.2	0.3	35.0	1.1	84.5
AGA preterm infants	12.7	0.6	8.5	0.3	35.0	1.1	84.7

*Area $= \frac{AP \times BP}{4} \times \pi$

At 40 postmenstrual weeks the difference of 1 cm (12.7 versus 11.7 cm) in the fronto-occipital diameter between preterm infants and full term infants is greatest. Subsequently this difference diminishes, at six months it amounts to 0.2 cm; by this the possible influence of head shape on head circumference becomes even more negligible. For significant changes of the area the difference in head length had to be considerably greater.

Velocity. From 30.5 to 39.5 postmenstrual weeks the growth velocity of head circumference is very high with a peak of 4.3 cm per month in the 34th week (Fig. 4 above).

After term velocity decreases rapidly. In the first month it amounts to 2.5 cm in the preterm infants, and is similar to that of the full term infants who grow 2.4 cm per month (Brandt 1978). In the sixth month, the velocity has slowed in both groups to 0.8 cm/month or to 20 %, and in the 18th month to 0.2 cm or 5 % of the mean peak velocity. At 4.5 years of age, the velocity decreases to 0.04 cm/month which is only one-hundredth of the peak rate (Brandt 1980b).

Full Term Infants with Phenylketonuria

Twenty-five early treated, i.e., during the first three months, infants with phenylketonuria (PKU) are followed prospectively in a longitudinal study. They were born between 1969 and 1981. Their growth and development is compared with that of a group of normal full term infants of the Bonn study (matched pairs).

Physical growth of the children with PKU does not differ from that of the normal controls. This is in accordance with recent findings in the literature, i.e. Holm et al. (1979). In fact, even controlling for family data such as social class, height of the father and mother, there appears to be a secular trend, i.e. some of the children grow at a higher percentile channel than would be expected for their target. In the past there has been often reported an unsatisfactory physical growth (Holm and Knox 1979).

SMALL-FOR-GESTATIONAL AGE (SGA) PRETERM INFANTS
PERMANENT GROWTH RETARDATION, AND CATCH-UP GROWTH PATTERNS

Nowadays there is an increasing interest in how infants with severe intrauterine growth retardation develop in later life. What possibilities does perinatal care offer today? Before the introduction of intensive care measures and of early and high-energy feeding a permanent growth retardation of head circumference and a reduced intellectual capacity usually was observed in SGA preterm infants. A catch-up growth of head circumference has not been reported except by Falkner (1966), Babson and Henderson (1974) and Brandt (1978).

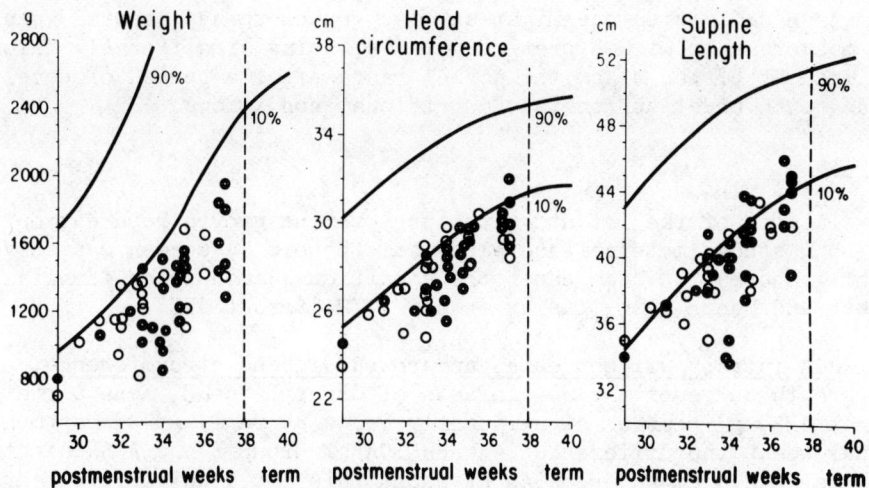

Fig. 8. Birth measurements of the 51 SGA preterm infants, plotted against the Lubchenco curves. Circles with an asterisk: SGA infants with catch-up growth, Group A, n=28; circles: SGA infants without catch-up growth, Group B, n=23.

Sample. Among the 115 preterm infants of the Bonn study 51 with a birthweight below the 10th percentile of Lubchenco are classified as SGA; the birthweight is in all but six cases 1500 g or below (Fig. 8).

Head circumference catch-up growth has been chosen here as a criterion for separation of the heterogenious SGA preterm infants into two groups: Group A, with catch-up growth of head circumference and with a normal neurological and mental development, n=28. Group B without catch-up growth of head circumference and with a less favorable development, n=23.

Out of the 28 infants in Group A, 22 were born after 1971, and the mothers of four of the six earlier-born infants have participated in a prospective study on pregnancy progress and child development of the Deutsche Forschungsgemeinschaft, and have taken advantage of intensive prenatal care.

The SGA infants in both groups (Fig. 8: Group A, circles with an asterisk; Group B, circles) at birth were mostly retarded not only in weight, but also in head circumference and supine length; figure 8 further shows that between Group A with catch-up growth, and Group B without catch-up, there is no difference in the degree of growth retardation.

As a definition one might say that catch-up has occured when the measurements of SGA preterm infants, being significantly below the mean at birth, reach the normal range after a period of accelerated growth under appropriate nutritional conditions.

Results

In most of the infants with intrauterine growth retardation in the Bonn study, maternal and placental factors have been causative (Tab. 1 above); in some cases the condition seemed to be familial (Scott and Usher 1966; Tanner et al. 1972; Brandt 1975).

SGA infants without catch-up growth of head circumference. The growth curve of the SGA infants of Group B (n=23, mean birthweight 1230 g) remains significantly below average. At 35 postmenstrual weeks the difference between SGA of Group B and AGA infants amounts to 2.3 cm and remains at about this level until five years of age (Brandt 1978).

The lack of postnatal catch-up growth of head circumference may be due to restrictive feeding practices in former years. Within the 10-year period of the Bonn study, started in 1967, there have been quantitative changes in immediate postnatal feeding of preterm infants. Most of the SGA infants in Group B (15 out of 23) were born before the introduction of early and high-energy feeding, i.e., 3-6 hours after birth, in July 1972. In the literature still today there is reported a permanent growth retardation of SGA infants (Holmes et al. 1977; Commey and Fitzhardinge 1979).

SGA infants with catch-up growth of head circumference. The amount of catch-up growth is difficult to judge, it can only be estimated by parameters as there is target head circumference. In general there are observed different patterns of catch-up growth, it can be complete or incomplete. A similar catch-up growth pattern in two different infants has to be judged differently in view of the different head sizes of their parents.

In 1973 in a group of SGA infants of the Bonn study a catch-up growth of head circumference became evident (Brandt and Schröder 1974). This catch-up was favoured by the introduction of early and high-energy feeding of preterm infants (Weber et al. 1976). In the meantime this first observation could be confirmed in further cases.

At 35 postmenstrual weeks the head circumference of the SGA infants of Group A differs significantly from that of the AGA infants by 2.1 cm ($p < 0.01$). Given early and high-energy feeding these SGA infants exhibit catch-up growth. The difference from the AGA control infants decreases to 0.5 cm at six months, and is no longer significant ($p > 0.05$). At five years the difference between both groups has decrased to 0.2 cm. All these SGA infants have shown a normal neurological, motor and mental development.

The favorable results may well be attributed to the close supervision of the mothers during pregnancy, and to the improved postnatal care of their offspring. Our observations suggest that catch-up growth in head circumference is only possible providing the infant receives early and high-energy feeding. On the other hand, the SGA infant, who looses weight after birth, and is not given the opportunity of early and high-energy feeding, shows no evidence of catch-up in head circumference in later life in spite of attempts of nutritional rehabilitation.

Also after discharge from the hospital nutrition of SGA infants must be appropriate in order to support further the process of catch-up growth; these SGA infants are still in danger to relapse into a growth retardation because their needs of energy are underestimated. Therefore advising the parents as to the optimal way of feeding and regular measurements of head circumference are important. In some SGA infants there are feeding difficulties, so that the mother needs a lot of patience. Further it is known that an environment which offers poor and inadequate stimulation may aggravate the disadvantageous effects of malnutrition. Therefore the SGA infant is especially in need of loving attentive care and stimulation. Changing caretakers in cases where the mother is practising a profession can be a great disadvantage for very small SGA infants.

<u>Conclusions.</u> The developmental prognosis of SGA infants involves many factors, acting during the pre-, peri- and postnatal period. From our results can be concluded that the development of SGA preterm infants of even very low birthweight can be supported in such a way that it takes a normal course by providing good environmental conditions, such as appropriate nutrition, especially during the immediate postnatal period with its high growth velocity, together with mental stimulation.

ACKNOWLEDGMENTS

This research was carried out with grants from the Minister für Wissenschaft und Forschung des Landes Nordrhein-Westfalen, from the Fritz Thyssen Stiftung, and from the Milupa A.G., International Scientific Department, which are gratefully acknowledged.

REFERENCES

Babson, S.G., Behrman, R.E., Lessel, R., 1970 Fetal growth: Liveborn birth weights for gestational age of white middle class infants, <u>Pediatrics,</u> 45:937.
Babson, S.G., Henderson, N.B., 1974, Fetal undergrowth:Relation of

head growth to later intellectual performance, Pediatrics, 53:890.
Battaglia, F., Lubchenco, L.O., 1967, A practical classification of newborn infants by weight and gestational age, J. Pediat., 71:159.
Baum, J.D., Searls, D., 1971a, Head shape and size of newborn infants, Develop. Med. Child Neurol., 13:572.
Baum, J.D., Searls, D., 1971b, Head shape and size of preterm low-birthweight infants, Develop. Med. Child Neurol., 13:576.
Brandt, I., Schröder, R., 1974, Postnataler Entwicklungsausgleich bei Frühgeborenen mit pränataler Dystrophie, Mschr. Kinderheilk., 122:697.
Brandt, I., 1975, Postnatale Entwicklung von Früh-Mangelgeborenen, Gynäkologe, 8:219.
Brandt, I., 1978, Growth dynamics of low-birth-weight infants with emphasis on the perinatal period, in: "Human Growth," Vol. 2, F. Falkner and J. M. Tanner, eds., Plenum Publishing Corporation, New York.
Brandt, I., 1979a, Perzentilkurven für das Kopfumfangswachstum bei Früh- und Reifgeborenen in den ersten sechs Jahren, der kinderarzt, 10:185.
Brandt, I., 1979b, Perzentilkurven für die Gewichtsentwicklung bei Früh- und Reifgeborenen in den ersten fünf Jahren, der kinderarzt, 10:713.
Brandt, I., 1980a, Perzentilkurven für das Längenwachstum bei Früh- und Reifgeborenen in den ersten fünf Jahren, der kinderarzt, 11:43.
Brandt, I., 1980b, Postnatal growth of preterm and full term infants, in: "Human Physical Growth and Maturation: Methodologies and Factors," F. Johnston, A. Roche, C. Susanne, eds., Plenum Press, New York and London.
Brandt, I., 1981, Kopfumfang und Gehirnentwicklung - Wachstumsretardierung bei intrauteriner Mangelversorgung und ihre Aufholmechanismen - , Klin. Wschr., 59:995.
Brenner, W.E., Edelman D.A., Hendricks, C.H., A standard of fetal growth for the United States of America, Amer. J. Obstet. Gynec., 126:555.
Commey, J. O. O., Fitzhardinge, P. M., 1979, Handicap in the preterm small-for-gestational-age infant, J. Pediat., 94:779.
Dobbing, J., and Sands, J., 1978, Head circumference, biparietal diameter, and brain growth in fetal and postnatal life, Early Hum. Dev., 2:81.
Dunn, P. M., 1979, Perinatal terminology, definitions and statistics, in:"Perinatal Medicine, Sixth European Congress, Vienna," O. Thalhammer, K. Baumgarten, and A. Pollack, eds., Georg Thieme Publishers, Stuttgart.
Falkner, F., 1966, General considerations in human development, in: Human Development, F. Falkner, ed., W.B. Saunders Company, Philadelphia, London.

Gruenwald, P., 1966, Growth of the human fetus. I. Normal growth and its variation, Amer. J. Obstet. Gynec., 94:1112.

Hagen, W., 1966, Wachstum und Entwicklung von Schulkindern. Bericht über die Längsschnittuntersuchungen in Deutschland, Dtsch. med. Wschr., 34:1490.

Hansmann, M., 1974, Kritische Bewertung der Leistungsfähigkeit der Ultraschalldiagnostik in der Geburtshilfe heute, Gynäkologe, 7:26.

Hansmann, M., 1976, Ultraschall-Biometrie im II. und III. Trimester der Schwangerschaft, Gynäkologe, 9:133.

Hansmann, M., Schuhmacher, H., Foebus, J., and Voigt, U., 1979, Ultraschallbiometrie der fetalen Scheitelsteißlänge in der ersten Schwangerschaftshälfte, Geburtsh. u. Frauenheilk., 39:656.

Hohenauer, L., 1973, Intrauterines Längen- und Gewichtswachstum, Pädiat. Pädol., 8:195.

Hohenauer, L., 1980, Intrauterine Wachstumskurven für den Deutschen Sprachraum, Z. Geburtsh. Perinat., 184:167.

Holm, V. A., and Knox, W. E., 1979, Physical growth in phenylketonuria: I. A retrospective study, Pediatrics, 63:694.

Holm, V. A., Kronmal, R. A., Williamson, M., and Roche, A. F., 1979, Physical growth in phenylketonuria: II. Growth of treated children in the PKU collaborative study from birth to 4 years of age, Pediatrics, 63:700.

Holmes, G. E., Miller, H. C., Hassanein, K., Lansky, S. B., and Goggin, J. E., 1977, Postnatal somatic growth in infants with atypical fetal growth patterns, Amer. J. Dis. Child., 131:1078.

Hosemann, H., 1949, Schwangerschaftsdauer und Neugeborenengewicht, Neugeborenengröße und Kopfumfang des Neugeborenen, Arch. Gynäk., 176:109, 123, 443.

Iffy, L., Jakobovits, A., Westlake, W., Wingate, M., Catenni, H., Kanofsky, P., and Menduke, H., 1975, Early intrauterine development: 1. The rate of growth of Caucasian embryos and fetuses between the 6th and 20th weeks of gestation, Pediatrics, 56:173.

Karlberg, P., Taranger, J., Engström, I., Lichtenstein, H., and Svennberg-Redegren,I., The somatic development of children in a Swedish Urban Community. A prospective longitudinal study, Acta paediat. scand., Suppl. 258:1, 1976.

Kloosterman, G. J., 1969, Over intra-uterine groei en de intra-uterine groei-curve, Maandschr. v. Kindergeneeskunde, 37:209.

Largo, R.H., Duc, G., 1977, Head growth and changes in head configuration in healthy preterm and term infants during the first six months of life, Helv. paediat. Acta, 32:431.

Lubchenco, L. O., Hansman, C., Dressler, M., and Boyd, E., 1963, Intrauterine growth as estimated from live born birth-weight data at 24 to 42 weeks of gestation, Pediatrics, 32:793.

McKeown, T., and Gibson, J. R., 1951, Observations on all births (23970) in Birmingham 1947. "IV. Premature birth", Brit. med. J., II:513.

Meredith, H.V., 1946, Physical growth from birth to two years: II. Head circumference. Part I. A review and synthesis of North American research on groups of infants, Child Develop., 17:1.

Milner, R. D. G., and Richards, B., 1974, An analysis of birth weight by gestational age of infants born in England and Wales, 1967 to 1971, J. Obstet. Gynaec. Brit. Cwlth, 81:956.

Reinken, L., Stolley, H., Droese, W.,and Oost van, G., 1980, Longitudinale Körperentwicklung gesunder Kinder. II. Größe, Gewicht, Hautfettfalten von Kindern im Alter von 1,5 bis 16 Jahren, Klin. Pädiat., 192:25.

Robinson, H. P., and Fleming, J. E. E., 1975, A critical evaluation of sonar "crown-rump-length" measurements, Brit. J. Obstet. Gynaec., 82:702.

Scott, K. E., and Usher, R., 1966, Fetal malnutrition: its incidence, causes, and effects, Amer. J. Obstet. Gynec., 94:951.

Susanne, C., 1982, Heretability of head circumference, Personal communication.

Streeter, G. L., 1920, Weight, sitting height, head size, foot lenth, and menstrual age of the human embryo, Contrib. Embryol. Carneg. Inst., 11:143.

Tanner, J. M., Whitehouse, R. H., and Takaishi, M., 1966, Standards from birth to maturity for height, weight, height velocity, and weight velocity: British children, 1965-I and -II, Arch. Dis. Childh., 41:454 and 613.

Tanner, J. M., Lejarraga, H. and Turner, G., 1972, Within-Family standards for birth-weight, Lancet, 2:193.

Tanner, J. M., and Whitehouse, R. H., 1982, "Atlas of Children's Growth. Normal Variation and Growth Disorders," Academic Press, London, New York.

Thomson, A. M., Billewicz, W. Z., and Hytten, F. E., 1968, The assessment of fetal growth, J. Obstet. Gynaec. Brit. Cwlth, 75:903.

Usher, R., and McLean, F.H., 1969, Intrauterine growth of live-born Caucasian infants at sea level: Standards obtained from measurements in 7 dimensions of infants born between 25 and 44 weeks of gestation. J. Pediat., 74:901.

Weaver, D. D., and Christian, J. C., 1980, Familial variation of head size and adjustment for parental head circumference. J. Pediat. 96:990.

Weber, H. P., Kowalewski, S., Gilje, A., Möllering, M., Schnaufer, I., and Fink, H., Unterschiedliche Calorienzufuhr bei 75 "low birth weights": Einfluß auf Gewichtszunahme, Serumeiweiß, Blutzucker und Serumbilirubin, Europ. J. Pediat., 122:207, 1976.

Wieringen van, J. C., Wafelbakker, F., Verbrugge, H. P., and Haas de, J.H., "Growth Diagrams 1965 Netherlands Second National Survey on 0-24-Years-Olds," Wolters-Noordhoff Publishing, Groningen, 1971.

MEDICAL ENVIRONMENT: APPLICATIONS IN THE FIELD OF PUBLIC HEALTH

Comments on the Report of I. Brandt

Geneviève Massé

Professor, National School of Public Health, France
Avenue du Professeur Léon Bernard
35043 Rennes Cedex (France)

I. INTRODUCTION

This comment on Dr. Brandt's paper was written, as requested, from the point of view of the public health epidemiologist.

Indeed, the problems posed by small-for-gestational-age infants are of major concern in public health for several reasons:

- The magnitude of its yearly world incidence: about 16,000,000 in the underdeveloped countries and 350,000 in the industrialized countries (Falkner, 1981).

- The bad prognosis, when adequate care of the newborn is not available.

- The high cost of adequate care, in terms of financial and economic cost for society as well as psychological stress for the families.

When concerned with perinatal problems, health services must deal with

1. The prevention of intra-uterine growth stunt (primary prevention)

2. If this growth stunt does appear, secondary prevention: prevention of death and prevention of adverse health conditions for the survivors.

This evidently cannot be achieved without pediatric research, and Dr. Brandt's rigorous study is of particular interest to public health.

Much of the thinking of the epidemiologist, and of the action in public health, is based on the notion of risk. In this respect, this paper on intra-uterine growth and its influence on postnatal growth brings up the following points:

1. The reference curves of growth, which should be describing the growth of low-risk infants and children in a representative reference population.

2. The limits to be used for these curves (sometimes called "limits of normality"), which should be a threshold from low to high risk; or the deviations from longitudinal patterns which can be considered as an indicator of risk.

3. The measurements which are most relevant as risk detectors.

4. The factors associated with deviations from "normal" growth, and among these the etiological factors whose knowledge permits primary prevention.

5. The management necessary to promote secondary prevention.

These five points will be kept in mind, in discussing Dr. Brandt's paper.

II. REFERENCE CURVES: THEIR REPRESENTATIVITY

Weight curves which have been used since Lubchenco, i.e., cross-sectional birthweights of prematurely born infants, have been and still are extremely useful, evidently, especially to distinguish small-for-gestational-age from adequate-for-gestational-age preterm infants, but what they represent must be kept clearly in mind:

1. Since they concern prematurely born infants, obviously one cannot assume a priori that they are representative of "normal" intra-uterine growth, that is, growth resulting in a low-risk newborn.

2. They are representative of prematures' birthweight in a particular population and a particular environment which must be both clearly defined. In the case of Lubchenco's data, the births occurred at the altitude of Denver, where the birthweight distribution is slightly shifted towards lower values (Lichty et al., 1958). This might account for weights lower than in the other studies cited in this report; but a number of other factors might intervene, such as the structure of the maternal population as regards age, parity, social status, and so on.

These differences in structure might also occur in one study, for different gestational ages: are the populations of mothers and of fetuses, for births occurring at different gestational ages, comparable? Do they have the same characteristics? For instance, is the age and social structure of women delivered at 28 weeks the same as women delivered at 38 weeks?

If not, we cannot assume that the population of fetuses born at 38 weeks had an intra-uterine weight distribution at 28 weeks similar to the birthweight distribution of fetuses born at 28 weeks. One should therefore be extremely cautious when deriving velocity curves from these cross-sectional data.

As regards the early-growth curves, based on weight of embryos after voluntary abortion, the same cautious approach is needed, since the population of women undergoing abortion might be different from the ones having live births. For instance, an unpublished study done in Brittany (France) has shown a greater proportion of adolescent girls and of women over 35.

The above remarks about international comparisons apply of course to post-natal curves; one can wonder whether the data here compared are representative of the countries where they were collected and belong to the same period; this is certainly not the case for the French study mentioned, which was not designed to be representative of the population, and which concerns children born some 20 years earlier than the children observed by Dr. Brandt.

III. LIMITS OF "NORMALITY"

When considering growth standards, the problem arises about the limits to be chosen (cross-sectional data) - in the present case the lower limit - or the patterns of growth (longitudinal data) which lead to a pathological situation.

In distance cross-sectional data, the limit should define a threshold under which the risks increase sharply: risk of death, risk of poor development.

For example, the limit of 2500 g for birthweights was based on studies - done in industrialized countries - of neonatal death rates in relation to birthweights; this threshold might be different in different parts of the world (Vergara-Valenzuela, 1969). It is presumably lower now in countries that are well equipped.

These limits of reference curves, however, are often chosen for purely mathematical reasons - for instance, the mean, 2SD (standard deviations seem to fascinate some pediatricians), or the 1st decile; one would often like to know why these limits are chosen, in relation to risk. For instance, in her definition of

SGA infants, the author has chosen the 1st decile of Lubchenco's weight curve, that is 44% of the prematures under study: does it represent a distinct threshold of risks for her own population of newborns?

As regards the "abnormal" patterns of growth, when longitudinal data are available one resorts to flattening of the distance curves, or a loss of velocity as compared to velocity distributions; the point of issue in this case might then be the length of time during which this flattening goes on; and it is this measure of time that could be a risk indicator, risk of inadequate catch-up growth later on, i.e., of irreversibility of pathological conditions.

The author of the report gives a nice discussion of what one should consider a pattern of adequate catch-up growth - the difficulty of assessing it for individuals, compared to the easier definition for a group: "it means that the results do not differ significantly from those of the control infants."

IV. THE CHOICE OF MEASUREMENTS

Of the dimensions that can be measured or estimated, which are the most useful as health (or, conversely, as risk) indicators?

The report mentions

> Weight
> Supine length and crown-rump length
> Head circumference
> Biparietal diameter
> Thoracic diameter

Weight

Birthweight is used here as indicator to differentiate, among preterm infants, those who are small for gestational age (SGA) from those who are adequate for gestational age (AGA), the first group being more at risk than the second. Weight is a global measurement, and as such very sensitive to change, which is an advantage; but as this report shows, other measurements are needed to define specific risks, which are discussed below. It cannot, of course, be measured prenatally; it is therefore an indicator for secondary prevention only.

Length

Apparently, this is not considered here as a good risk indicator for infants, since at 40 postmenstrual weeks, the length of infants considered adequate for gestational age is "by 2.8 cm significantly less than full-term controls."

MEDICAL ENVIRONMENT

Crown-rump length

This measurement, which can be done by ultrasonic technique in the first half of pregnancy, is presented here, when taken longitudinally and translated into velocity, as a "very accurate means of assessing maturity in the first trimester of pregnancy." Does it imply that the occurrence of early growth retardation is rare?

Standardization of techniques for interstudy comparisons seems as yet to be inadequate, as shown by the very important differences between Hansmann's and Robinson's results.

Head circumference

This measurement can be used to estimate brain weight.

Its pattern of growth, here catch-up growth, is used as a specific indicator to differentiate, among SGA defined by the weight indicator, those at risk of poor brain development; it is of course of extreme importance in secondary prevention; it is discussed in point V of this comment.

Biparietal diameter

The great advantage of this measurement is that it can be done during intra-uterine life. However, brain weight cannot be estimated from it. Is it, despite this fact, a good indicator of brain development?

In this respect, the curves describing prenatal growth of this measurement in SGA infants are very striking; here again it is the flattening of the curve which is an indicator

The timing of this flattening which occurs always after 28 weeks does not seem to be the same for all infants who belong to "group A," that is, who have experienced catch-up growth after birth.

It would have been interesting to compare these curves with those of group B infants, who did not catch up after birth, to see whether the timing and duration of biparietal diameter stunt was different from the other group: it has been suggested (Falkner, 1976) that timing and length of malnutrition might be of vital significance for the growth of the brain.

Thoracic transverse diameter

This measurement is stated to be an indicator of liver development. Further studies on the consequences in later life of the slowing down here shown would be of great interest.

It is stated also that in 70% of the cases of growth retardation the trunk is affected before the other parts of the body; this diameter should therefore be a sensitive indicator of intra-uterine growth retardation.

V. PREVENTION

1. <u>Primary prevention</u> (action against etiological factors)

This subject is briefly mentioned in the report, which supplies a list of the etiological factors.

It would have been interesting to know what specific actions have been taken in Dr. Brandt's service to prevent the occurrence of intra-uterine growth retardation.

Until recently, it seems that — as multivariate statistical analysis has shown (Goujard et al., 1974) — even when taking into account all known risk factors of low birthweight (which are not necessarily causative), the predictive possibilities were poor. In other words, one could not detect with great accuracy, early in pregnancy, the actual amount of "risk" to be overcome. The very good results obtained in recent years concerning the outcome of pregnancies is a very positive fact. In achieving this, however, it is possible that more care than would have been necessary has been delivered to women who were in fact low-risk women and were treated as "high-risks," causing unnecessary cost and unnecessary annoyance for the women concerned.

With the development of new techniques, were more specific actions possible?

International comparisons between countries of equal economic level show the importance of the social laws and customs protecting the pregnant woman (Alberman, 1980). When it comes to individual cases, however, it is difficult, early in pregnancy, to estimate the risk of low birthweight to which each women is submitted, and therefore to take adequate decisions.

For instance, Goujard et al. calculated risk coefficients by Feldstein's method using a multiple regression function; they used the 17 variables most closely related to low birthweight to classify the women early in pregnancy, in relation to the risk of LBW. In a population of women in which the overall risk of low birthweight was 4.4%, this method had a positive predicting power (probability of having a LBW child, when classified "moderate" or "high" risk) of only 7.7%, and a negative predicting power (probability of having a child of birthweight larger than 2500 g, when classified as "low risk") of 97.5%. This classification therefore, while requiring a long questionnaire, does not help much in decision making.

MEDICAL ENVIRONMENT

2. Secondary prevention

Secondary prevention is concerned with the management of small infants in order to achieve

- prevention of death
- prevention of poor development in the survivors.

The report deals only with the second point: were there any deaths of SGA during the period of observation, and, if so, was any relationship found between intra-uterine growth of the measurements here studied (such as thoracic diameter) and the risk of death?

This report brings very satisfactory and important results concerning the management of SGA to promote normal postnatal psycho-neuromotor development: it shows the significantly better results obtained since the start of a new program in 1972. Some questions are brought to mind:

a) What are the limiting factors likely to prevent the catch-up growth of the head? (Among the 30 children who were taken care of after 1972, 8 did not show catch-up growth of the head. In what way were these children different from the 22 others who apparently underwent the same type of care? Can limiting factors be detected and measured?)

b) What is the relationship between brain quantitative growth, estimated from head circumference growth, and the development of mental capacities? (Dobbing, 1976).

It is shown that all children with catch-up growth experienced a normal neuromotor and mental development, but no details are given about the children without catch-up growth: did any of these develop satisfactorily, and if so, was their familial environment different from the others? For that matter, was the environment and care for the children with catch-up growth different from that for those without?

Considering the magnitude of SGA births in the developing world, this last point is essential. In traditional societies, mother's care and environment are very adequate during the first months, as shown by the amazingly good psychomotor development of the infants (Massé, 1969). To what degree does this make up for stunted intra-uterine brain growth?

Where important national perinatal programs have been set up, perinatal mortality has dropped drastically. Although other factors have played a part in this drop (for instance, wider use of contraception), there is no doubt of the efficacy of these policies. More controversial has been the prevention of handicaps: was the saving of very-low-birthweight babies creating handicaps?

Stewart et al. (1981) describe four stages of the perinatal situation:

1. In the 1950's, little attention was paid to very low birthweight, and death rate was high.

2. During the next decade, there was a better prevention of death, but sometimes at the expense of the health of the infant (for example, the occurrence of retrolental fibroplasia).

3. Then, until recently, the death rate diminished, with the rate of handicaps remaining constant.

4. There is probably now a decrease in both rates, with the setting up of programs to monitor morbidity as well as mortality. However, the evaluation of the incidence of handicaps is much more difficult than the measure of mortality; one ought to use very precise and standardized criteria, which is not always the case; one should define also the ages at which handicaps are observed since they might take several years to become apparent.

VI. CONCLUSION

More epidemiological investigation is needed, using rigorous medical investigation techniques such as those presented in this report, to end up with still more efficient prevention policies. In this respect, international investigation is a great asset provided all known environmental and individual factors are rigorously controlled and techniques of measurement and data recording standardized.

To the public health administrator, key issues remain, even as knowledge of what better prenatal and postnatal care should be progresses.

- In our industrialized countries, are all women in need of this care, and especially high-risk women, actually getting it?

- In developing countries, how can care be adapted to local conditions and habits, which often are very beneficial? (Sanokho, 1980).

Multidisciplinary investigations and actions are in progress to answer these challenges.

REFERENCES

Albermann, E., 1980, Prospects for better perinatal health, The Lancet, 189-192.

Dobbing, J., 1976, Malnutrition et développement du cerveau, La Recherche, no. 64, vol. 7:139.

Falkner, F., 1976, Application and approach in community work: needed research, in at risk factors affecting health and nutrition of young children, Cairo. Supreme council for population and family planning, Cairo.

Falkner, F., 1981, Maternal nutrition and fetal growth, Am. J. Clin. Nutrition, 34:769.

Geber, M., 1973, L'environnement et le développement des enfants africains, Enfance, 3-4:145.

Geber, M., 1977, Le développement des nouveaux-nés africains, Courrier du C.I.E., no. 5, vol. XXVII:425.

Goujard, J., et al., 1974, Prévision de la prématurité et du poids de naissance en début de grossesse, J. Gyn. Obst. Biol. Repr., 3:45.

Lichty, J.A., et al., 1958, Studies of babies born at high altitude. I. Relation of altitude to birthweight, A.M.A.J. Dis. Chldr., no. 6, vol. 93:666.

Massé, G., 1969, Croissance et développement de l'enfant à Dakar, Monographie du C.I.E., Paris.

Sanokho, A., 1980, Obstétricie sociale, Courrier du C.I.E., no. 4, vol. XXX:345.

Stewart, A.L., Reynolds, E.O.R., and Lipscomb, A.P., 1981, Outcome for infants of very low birthweight: survey of the world literature, The Lancet, 1038.

Vergara-Valenzuela, A., et al., 1969, Relationship of birthweights and survival rates of Filipino infants born in Manila. J.P.M.A., July 1969.

LINEAR AND CATCH-UP GROWTH IN CHRONICALLY ILL CHILDREN

L. Greco and M. Mayer-Greco

Marina University of Naples
Department of Pediatrics
II School of Medicine, Naples, Italy

Professor Brandt's report proposes a number of topics of great interest to the neonatologist and the pediatrician in general. My comments will try to take from the specific subject of growth at the early ages those topics and suggestions that may be widely extended to other chronic pathological conditions.

The growth of SGA babies may well be taken as a model for other chronic conditions which appear at the later ages: distance, velocities, growth of different segments, catch up growth: all may be well used to follow the evolution of the disease, the remission, the quality and efficacy of treatment, the complicance of the patient.

GROWTH CURVE IN CHRONIC PATIENTS

Most chronic diseases of infancy and childhood affect growth to a considerable degree, so that it is possible to use growth evaluation as a reliable marker of the evolution or remission of the basic condition. But this is really possible only when growth is followed up at regular intervals, since it is the individual growth curve, and velocities, which are most informative to the pediatrician.

When a chronically ill child is subjected to clinical assessment, an accurate evaluation of his growth should be made, with the precision required by short time intervals and the need to pick up minimal variations in his growth pattern.

Figure 1 shows the growth curves (dots) in height and weight of a child affected by Cystic Fibrosis: his stature is consistently stable, but his weight is not as good, due to the severe pancreatic deficiency. This child had been diagnosed at 6 months of age, and a very active home treatment had been established in the first year of life. The squares show a similar growth curve for a girl affected by the same disease. Unfortunately the girl died at 9 years, having suffered from severe respiratory disorders from the age of 4. The family provided adequate treatment until the age of 4, when the mother refused to take care of the child any longer. From that time on the growth curve showed a continuous fall. Her growth velocity shifted toward lower centiles, and her height and weight Standard Deviation score (z-score) was progressively smaller (-2 to less than -4).

Figure 2 shows the growth pattern of a typical coeliac patient. The course of his weight curve may depict the different phases of his disease: the fall before the diagnosis, the catch-up, a second relapse due to the gluten challenge, and the final catch-up to a stable growth. The height curve is a bit less informative since there were hardly any height data before diagnosis, and the challenge is not so severe as to affect height growth significantly.

These practical examples show the immediate utility of growth curves in different chronic conditions: the message is peculiar for each of them, so that in different cases distance, or velocities, or shape of the growth curve may illustrate the condition of the patient better than a single clinical evaluation.

CATCH-UP GROWTH: THE CASE OF COELIAC CHILDREN

Discussion of Old Questions

Prader, Tanner and Von Harnack's discussion in the celebrated paper of 1963 received lots of comment but few answers, probably because of the difficulty of studying groups of sick children of adequate size longitudinally, with an accurate eye to the auxologists' technique.

At the Pediatric Department of Naples University we have followed 261 coeliac children for an average time of 3.7 years, recording their growth data at each visit. We now have preliminary data of a cross-sectional nature on their growth, but more time is needed for a mixed longitudinal analysis.

Still, it is possible to get from the mean increments in height and weight after the diagnosis some interesting suggestions about the factors that may modulate catch-up growth.

Data on catch-up growth are presented pooled for the two sexes since no significant difference between the sexes appeared in any phase of the analysis.

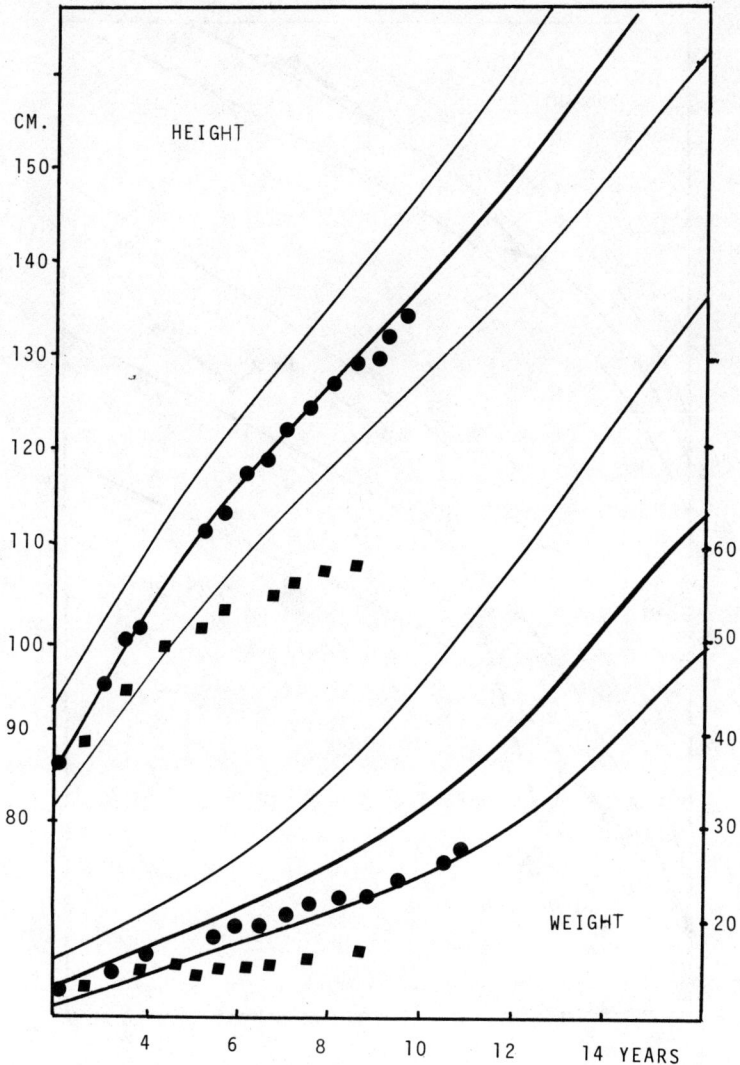
Fig. 1: Children affected by cystic fibrosis.

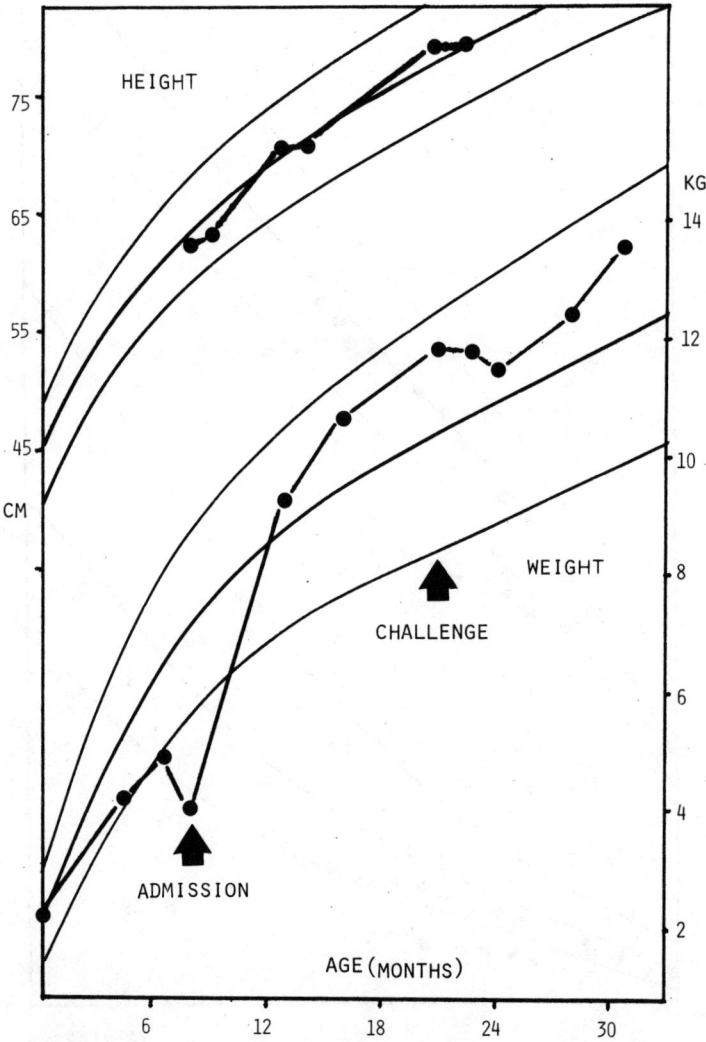

Fig. 2: Growth pattern of a coeliac patient.

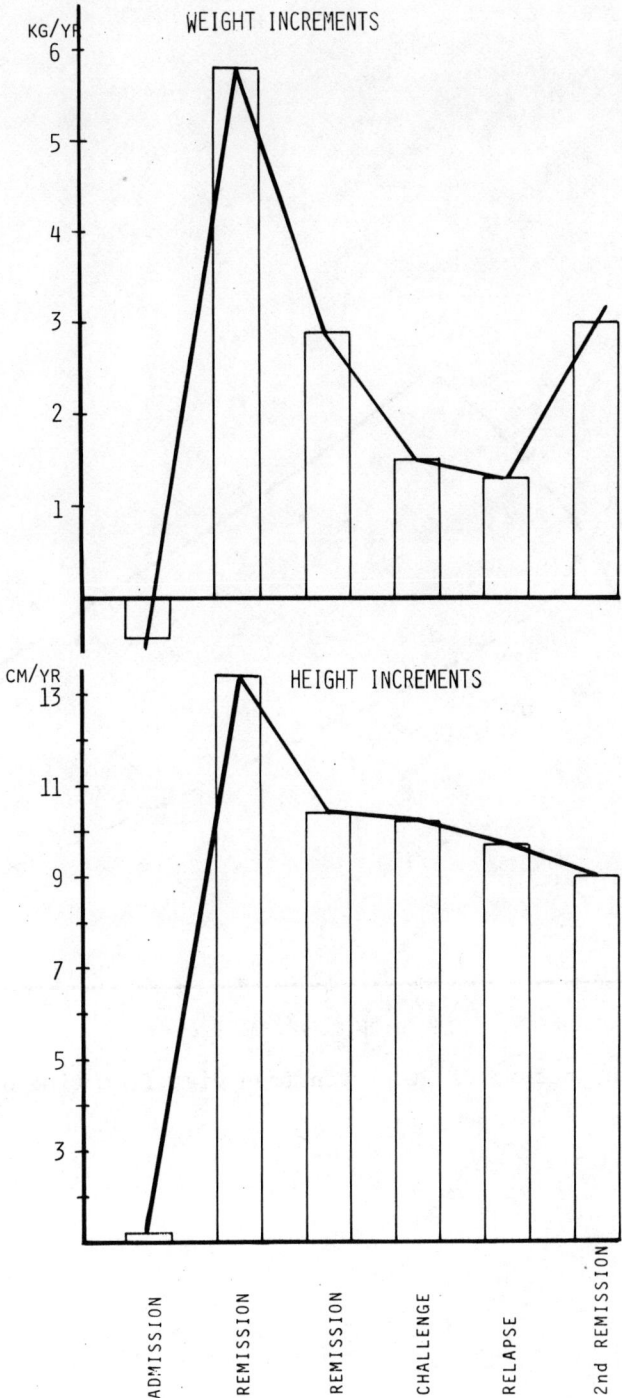

Fig. 3: Weight and height increments in coeliac patients.

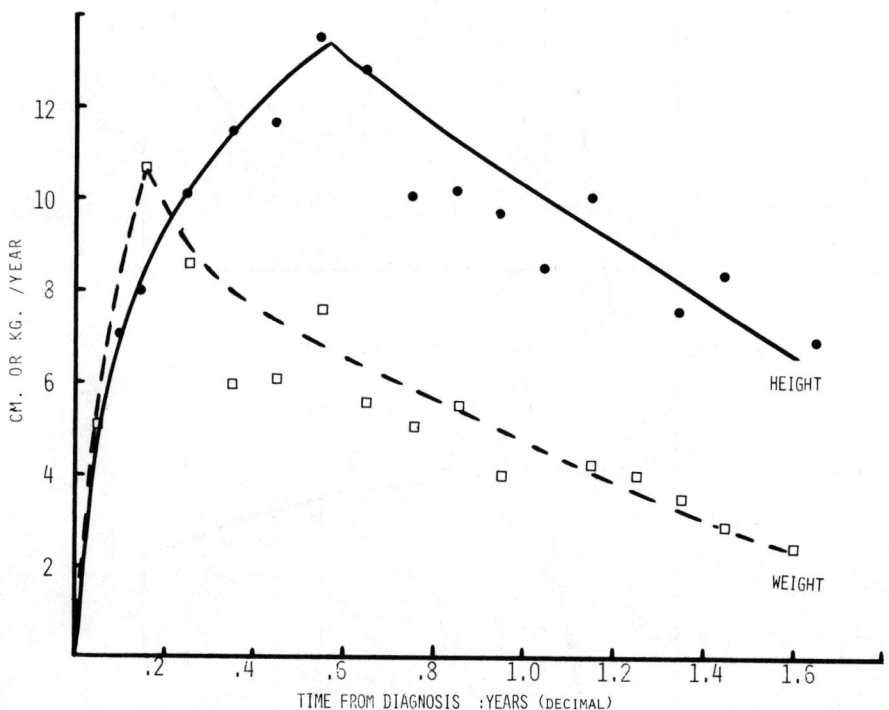

Fig. 4: Mean increments at gluten free diet from time of diagnosis.

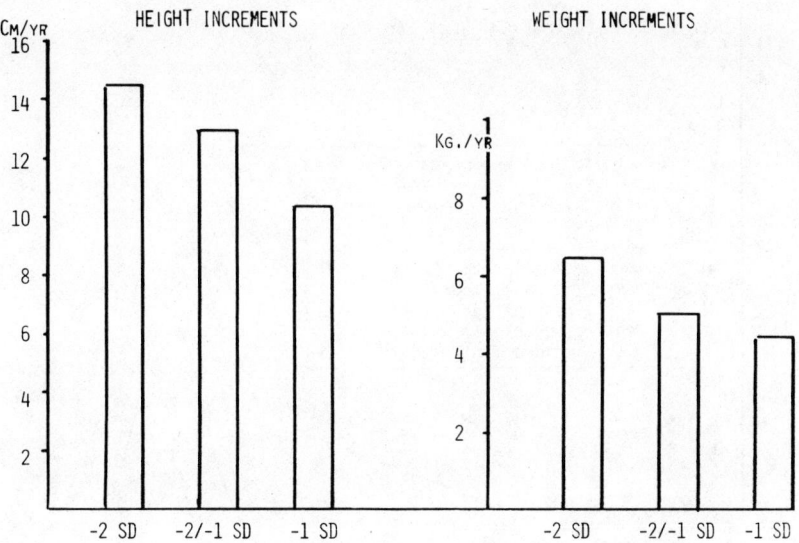

Fig. 5: Standard deviation score at admission.

Mean increments (kg/yr or cm/yr) during the different phases of the disease are given in Figure 3. Increments in weight are maximal soon after the diagnosis is made, slow progressively during the "remission" at gluten free diet, still go down after reexposure at gluten-containing diet and show a second catch up as soon as the gluten is again taken off the diet. Height increments show also a sudden catch up, which appears to be more "diluted" in time as compared to weight. The glutten challenge is not severe enough to show any steeper fall in height increments or a second catch up.

Catch up growth appears from Figure 4 to be much earlier for weight than for height, when increments are plotted against time from diagnosis, and therapy. While mean weight increments have their maximum at about 2 months from 0 time (diagnosis), and may well be earlier, maximum height increments are not reached before 6 months from 0 time.

Another question from Prader was about the possible relation between the severity of the growth failure and the amount of catch up in the individual child: Figure 5 seems to show that this relation holds for our patients: the more depressed is the Z-score at

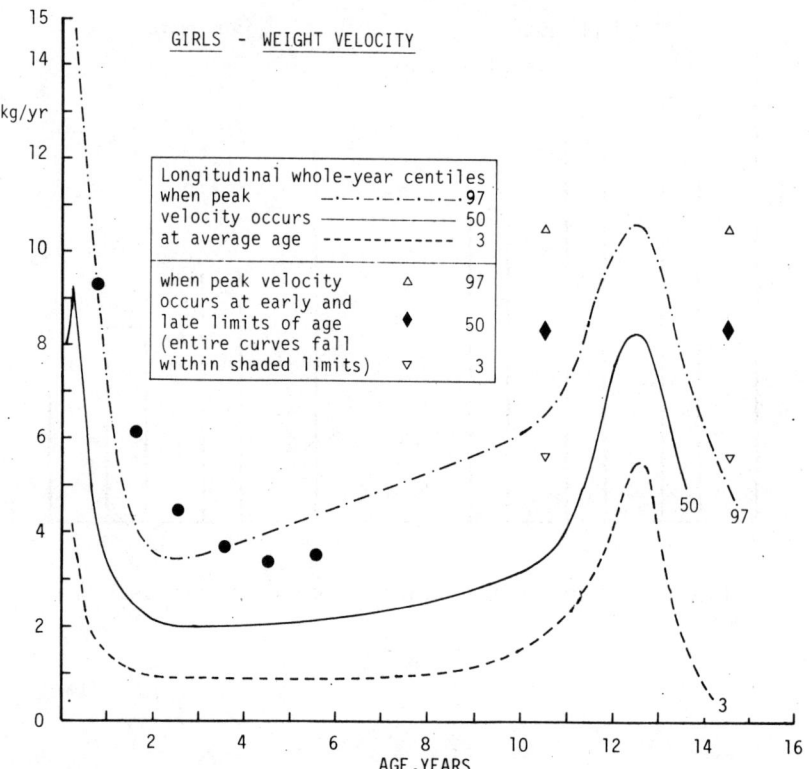

Fig. 6: Mean increments during early catch-up: weight.

time of diagnosis the higher is the mean increments in weight as well as in height. In 1963 Prader suggested that maximum catch up velocity could have been related to the actual age of the subject, but concluded that "our case prove no direct evidence in favour of this view". Possibly the present discussion on coeliac children gives some evidence in favour of this view. Indeed our suggestion is that the major factor which affects the size and possibly the timing of the catch up is actually the age of the individual subject.

Figure 6 illustrates the mean increments in the "remission phase" for weight plotted against age: it is quite evident that the higher the age the smaller the increments, with a pattern which is remarkably parallel to the ordinary velocity curve. A similar pattern is observed for height (Figure 7), where it has to be considered that, generally

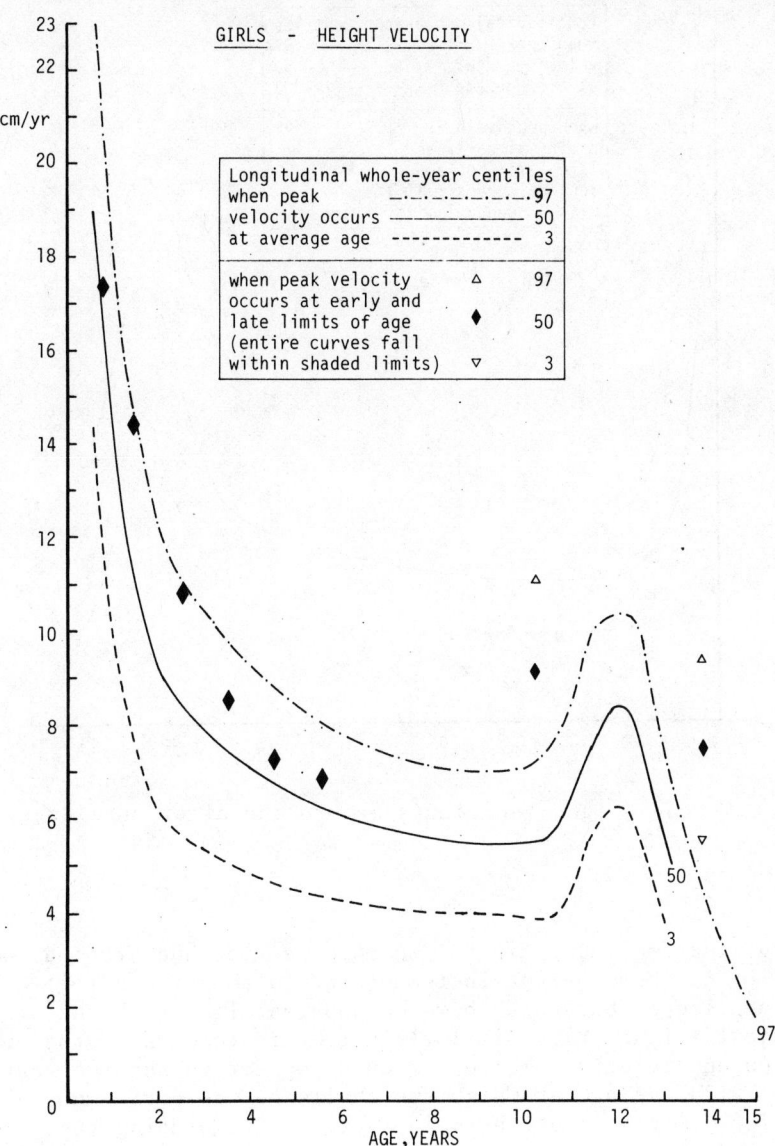

Fig. 7: Mean increments during early catch-up: height.

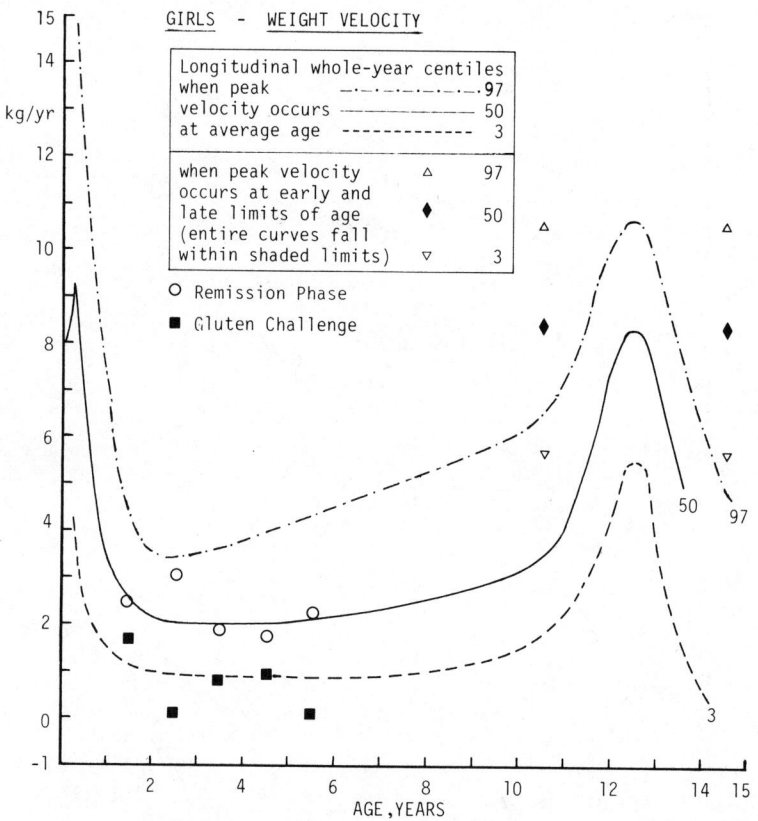

Fig. 8: Mean weight increments before and after challenge.

speaking, the coeliac patients are more wasted than stunted, before the age of 6. Mean weight increments are also quite affected by the gluten challenge, but this shows no apparent link with the age of the subject (Figure 8). Similarly weight increments at the second remission and relative smaller catch up appear to show no statistical relation with the age at which the challenge was performed. This "Age effect" may well be the main factor in regulating the catch up, but, in our case, it is not likely that younger children (who come earlier to treatment) are more wasted: so the effect due to the degree of growth retardation remains quite probable. Obviously only the results from the longitudinal analysis will give better answers to these problems, specially because of the strong age effects on the catch up growth.

In conclusion, the pediatrician who deals with chronic patients has a lot of auxology to plan and to do: the coeliac disease is a stimulating model, due to the relative homogeneity of the growth pattern, but in most other conditions useful information will be derived from the analysis of growth curves of individuals and of groups.

REFERNCES

1. A. Prader, J. M. Tanner and G. A. von Harnack: Catch up growth following illness or starvation. J. Pediatrics, 62, 646, 1963.
2. D. G. Barr, D. H. Shmerling and A. Prader: Catch up growth in malnutrition, studied in Celiac Disease after institution of a gluten-free diet. Pediat. Res., 6, 521-527, 1972.
3. J. Rey, F. Rey, J. Jos and S. Amusquivar Lora: Etude de la croissance dans 50 cas de maladie coeliaque. Arch. Franc. Ped., 28, 37-47, 1971.
4. D. P. Davies: Growth of small date babies. Early Human Development, 5, 95-105, 1981.

MEDICAL ENVIRONMENT - INFLUENCES ON GROWTH

Olcay Neyzi

Faculty of Medicine

Istanbul, Turkey

I. INTRAUTERINE GROWTH AND BIRTH WEIGHT

Dr. Brandt, in her report based on data from her extensive work on the longitudinal follow up of AGA and SGA preterm infants, beautifully demonstrates that in an optimal medical environment, the AGA preterm infant shows no deviation from normal growth and that the differences in measurement between corrected and uncorrected age become nil over the postnatal months. Dr. Brandt's data also show that, again given an optimal medical environment-- namely, early prenatal diagnosis, prenatal follow up, early delivery when indicated and postnatal care including meticulous perinatal care, early high energy feedings, adequate love and stimulation-- more than half of the preterm SGA infants showed a catch up in growth velocity, catch up in head growth being the earliest sign of a good prognosis.

Indeed, in areas of the world able to provide this care, intensive antepartum and intrapartum surveillance and improved perinatal care has resulted not only in dramatic decreases in neonatal mortality, including mortality rates of very low birth babies, but also in improved long term prognosis of AGA preterm babies (Stewart et al 1977, Creasy and Resnik 1981, Usher 1981, Svenningsen et al 1982). The same cannot be said about the long term prognosis of SGA infants. Attained physical development and incidence of neurologic sequelae in SGA infants vary in different series (Babson 1970, Drillien 1970, Fitzhardinge and Steven 1972, Gatson et al 1972, Fancourt et al 1976, Howard et al 1976, Low et al 1978, Holmes et al 1977). The overall results tend to show that, for those infants in whom the etiology of intrauterine growth retardation lies in late gestation and is relatively mild, the outlook is better.

Since this group is not represented in her series, Dr. Brandt does not dwell upon the full term SGA infant in her paper. It has been documented that the full term SGA infant is also at risk for perinatal complications such as intraventricular hemorrhage (Pape and Fitzhardinge 1981). Although the incidence of major neurologic sequelae is smaller in this group as compared to preterm SGA infants, a retrospective analysis of the pregnancy and birth weight history of 560 Swedish cases of cerebral palsy in the years 1950 to 1970 showed a disproportionately high representation of intrauterine growth retardation in term infants (Hagberg et al 1976). A review of the files of the Developmental Neurology Unit of our hospital in Istanbul, which also serves a center for handicapped children, has shown that, of the 104 cases followed in this unit for which birth weights and gestational histories were known, 32.4 % were full-term SGA infants.

Fetal malnutrition constitutes one of the major risk factors for stunted growth as well as for protein energy malnutrition, particularly in communities where perinatal care facilities are suboptimal and where nutritional requirements in early infancy are not met adequately. In the Endocrine Clinic of our hospital, 34 % of the non-hormonal cases of stunted growth were full term SGA infants. The degree of stunting in these cases was more exaggerated as compared to patients who were born at term with normal birth weights. An evaluation of cases of moderate to severe infantile PEM in a Maternal Child Health Center affiliated with our hospital also revealed that over half of the cases were term infants with birth weights lower then 2800 g (Neyzi et al 1979).

Full term SGA infants constitute a relatively high proportion of low birth weight infants, particularly among low socioeconomic groups and in nonindustrialized societies (Metcoff 1978). Studies in the United States and Great Britain show that about one third of all infants whose birth weight is less than 2500 g are not truly premature (Gruenwald 1964, Walker 1967). Van den Berg and Yerushalmy state that the gestation period is more than 37 weeks in a third to a half of low birth weight babies (Van den Berg and Yerashalmy 1966). According to a WHO publication, the proportion of LBW infants who are full term and small for gestational age ranges from 24 % to 57 % in developed countries (WHO 1980). In the same report, the proportion of LBW infants of 37 or more weeks gestation is given as between 34 % and 56 % in Africa, as 83 % in rural Guatemala, 38 % in Cuba and 75 % in India. A recent report suggests that in developing countries eight in ten LBW infants are born at term (Mata et al 1981). It is obvious that some of the variance in these figures is due to population differences in

birth weight and indicates the need for reliable data on birth
weights and intrauterine growth in different populations.

Knowledge about the intrauterine growth of infants of different
racial and ethnic groups is scarce. Similar to the construction of
postnatal growth curves, for an intrauterine growth to be useful for
detecting cases of deviant fetal growth in a community, it should
be derived from sections of that community living in optimal
conditions. Growth standards derived from the whole population
will often lead to an underestimation of LBW and small for date
infants, especially when there is a high incidence of maternal
malnutrition in the community (Olowe 1981). It is known that
duration of gestation obtained by history at time of delivery is
subject to error (Harrison and Roberts 1977, Hertz et al 1978).
The reliability the LMP history is enhanced when the patient is
seen early in gestation and the method is complemented by ultrasonic
evaluation of the fetus at this time (Varma 1973, Campbell 1974).
With these drawbacks in mind, a consideration of intrauterine
curves derived cross sectionally from gestational history at time
of delivery from mothers of low and middle socioeconomic status in
urban populations in Turkey (Ankara and Istanbul) reveal that
intrauterine growth in these populations compares well with standard
curves given in Dr. Brandt's report (Ozalp 1980, Erdogan, Neyzi,
Günöz 1982).

Of 2234 newborn-mother pairs included in a study on newborns
delivered in three large maternity hospitals in Istanbul, a reliable
history of LMP could be obtained in 1300 of the mothers. Analysis
of the birth weight data of infants born to these mothers also
showed that the intrauterine growth of the Turkish infant compared
well with North American and western European intrauterine growth
curves (Thomson 1968, Babson 1970, Brandt 1978). In Tables 1 and
2 are given the socioeconomic and anthropometric characteristics
of the mother-infant pairs in this study on newborns in Istanbul
city (Neyzi et al 1979).

These data show that birth weight and incidence of LBW in
Istanbul infants of low middle socioeconomic classes compare well
with figures reported for developed countries (Eveleth 1979,
WHO 1980). 52 % of LBW infants in the group were full term SGA
infants. Significant correlations were established in this group
between birth weight and gestational age, birth rank, placental
weight, mid upper arm circumference of the newborn and maternal
nutritional state assessed by relative weight.

To procure information on anthropometric and other data, 629
mother-infant pairs were investigated in obstetric hospitals in
four regions in Turkey outside if Istanbul (Neyzi et al 1982).
The populations in the regions selected for this study were a
mixture of small town and urban, and represent different ethnic

Table 1. Characteristics of the mothers

Mean age	23.8 ± 4.9 yrs
Proportion of mothers of low socioeconomic class	83.5 %
Proportion of illiteracy	21.5 %
Prepregnancy weight (by history) (mean ± SD)	56.7 ± 8.5 kg (n = 1113)
Predelivery weight (measured) (mean ± SD)	66.3 ± 8.7 kg
Postdelivery weight (measured) (mean ± SD)	60.9 ± 6.1 kg
Stature (measured) (mean ± SD)	156.7 ± 5.9 cm
Mid upper arm circumference (mean ± SD)	25.4 ± 2.5 cm
Proportion of underweight mothers (rel.wt < 90 %)	15.0 %
Proportion of short mothers (< 148.5 cm)	8.6 %

Table 2. Characteristics of the newborns

Number of boys	1142 (51.1 %)
Number of girls	1092 (48.9 %)
Gestational age (LMP) (mean ± SD)	39.5 ± 2.1 wks (n=1300)
Birth weight (mean ± SD)	3337 ± 542 g
Birth weight (first borns) (mean ± SD)	3254 ± 443
C-H length (mean ± SD)	50.0 ± 2.2 cm
Head circumference (mean ± SD)	34.6 ± 1.6 cm
Mid upper arm circ. (mean ± SD)	11.2 ± 1.1 cm
Proportion of LBW	6.5 %
Weight of placenta (mean ± SD)	596 ± 115 g
Weight of placenta (first borns) (mean ± SD)	573 ± 120 g

origin groups. All regions except Van, which is at an altitude of 2000 meters, are situated at or near sea level. Data from this study as compared to high socioeconomic and low middle socioeconomic class Istanbul children are given in Table 3 and show that with the exception of Adapazari, which is a relatively well off small town composed mainly of a population of circasian descent, mean birth weights were lower and the proportion of LBW infants higher in these regions. Comparison of the birth weights in different regions by birth rank and sex revealed parallel results.

The eastern Black Sea coast of Turkey and the eastern and south-eastern regions are areas showing the highest birth rates and highest infant mortality rates in Turkey. The educational level of the women is also particularly low in these regions. Medical facilities are also far from being optimal. Despite these realities, mean birth weight in the Black Sea group was only 100 g lower than the low middle class Istanbul newborns. The majority of the LBW babies were AGA preterm babies. Only 11.2 % of the mothers in this group were underweight as assessed by ponderal index values, showing that calorie needs were met in the majority of the mothers. Consequently, despite multiple pregnancies and low social class, intrauterine growth retardation did not appear to be a major problem in this region.

The lower birth weight encountered in the Van group is probably an outcome of adverse environmental influences enhanced by the possible effect of altitude. Maternal undernutrition, assessed by ponderal index, was encountered in 24.2 % of the Van mothers. 12.3 % of the Van, and 13.3 % of the Trabzon mothers were overweight by ponderal index evaluation. No significant differences were found between birth weights of infants born to underweight versus average weight mothers in either region. Infants of overweight mothers were significantly heavier. These results are in accordance with the findings reported by Ounsted (Ounsted and Scott 1981).

Birth weights were consistently low in infants born in Aydin, which is a small town in the Aegean coast hinterland. This population lives in relatively better socioeconomic conditions as judged by better housing, more spaced births, smaller proportion of illiterate mothers, fewer adolescent pregnancies, lower infant mortality rates as compared to the Trabzon (Black Sea) and Van (Eastern) regions of Turkey. Postdelivery weights of the mothers in the Aydin region and their calculated ponderal index values were significantly low compared to Istanbul, Adapazari, and Trabzon mothers. The preponderance of SGA babies in this community appears to be linked to the leanness of the mothers. This could be a genetic character or an outcome of dietary habits. Although these findings require confirmation by a larger series of mother-infant pairs and detailed information on maternal diet, it may be that optimal birth weight in the Aydin region is different from other

Table 3. Anthropometric and gestational data in Turkish newborns from different regions

Regions	n		Birth weight (g)	Head circ (cm)	C-H Length (cm)	Gestational age (weeks)		(%) proportion of		
						LMP	Dubowitz	LBW infants (<2500 g)	Fullterm SGA/LBW	Preterm infants GA<38 weeks (LMP)
ISTANBUL										
High socio-economic	B:	509	3414±575	–	–	–	–	–	–	–
	G:	574	3317±535							
Middle and low s.e. (from records)	B:	5753	3373±560	–	–	–	–	–	–	–
	G:	5337	3248±519							
(measured)	M:	2234	3337±542	34.6±1.6	50.0±2.2	39.6±2.1	–	6.5	52.0 (LMP)	12.2
OTHER (rural and small town)										
Adapazari	M:	163	3393±489	35.0±1.3	50.2±0.8	39.4±2.0	38.2±1.1	5.4	44.4 (Dubowitz)	20.9 (n:91)
Aydin	M:	111	3077±537	34.9±1.4	49.8±2.3	39.6±1.6	39.1±1.1	18.0	80.0 (Dubowitz)	8.1 (n:74)
Trabzon	M:	196	3234±489	34.6±1.5	49.7±1.9	39.4±2.1	39.0±1.5	8.5	11.8 (Dubowitz)	14.5 (n:90)
Van	M:	151	3127±450	34.6±1.2	49.6±2.4	38.8±2.6	38.4±1.7	11.8	35.3 (Dubowitz)	28.6 (n:14)

B: Boys,   G: Girls,   M: Both sexes

regions in Turkey, pointing to the need of local intrauterine growth curves and birth weight data not only for each country, bur for different communities within a country.

II. POSTNATAL GROWTH

Although genetic variations in body build exist among different populations, it is known that postnatal growth curves of high socioeconomic class children in many developing countries are almost identical with West European and North American standards (Eveleth 1978, Neyzi, Binyildiz, Alp 1979). It is also known that significant differences in level of growth and development are observed between children of different socioeconomic backgrounds in almost all countries (Tanner 1966, Neyzi et al 1973, Garn and Clark 1975, Neyzi et al 1975, Eveleth 1978). These differences have become minimal and nonsignificant only in some limited parts of the world (Goldstein 1971, Lindgren 1976, Eveleth 1978). The gap between children from different backgrounds becomes striking in developing regions where standard pediatric individual care is almost exclusive to the upper classes. Significant differences in height, weight, head circumference, age of attainment of puberty and intellectual performance are noted between high and low socioeconomic groups in developing regions, the differences being most striking when city children of high socioeconomic level are compared with children from rural areas (Gürson and Neyzi 1966). Our findings on Istanbul city children indicate that even relatively small differences in socioeconomic level influence growth and development (Neyzi et al 1973, Neyzi et al 1975). These differences result from inequities in economic, psychosocial, cultural, geographic, political and medical environment to which the growing organism is exposed.

The significant increase in adult stature and the trend towards earlier maturation noted in many countries during the past decades, the so-called secular trend, is an environmental phenomenon (Tanner 1966, Van Wieringen 1978).

Figure 1 demonstrates that secular trend continues to occur in high socioeconomic class children in Istanbul. The original Turkish growth standards were based on data from children of high socioeconomic level born between the years 1950 and 1960 and followed longitudinally in the first 8 years of life. Values on chronological ages 9 to 18 years were compiled from cross-sectional data on school children (Neyzi 1979). The higher values given in Figure 1 were also obtained from the files of high socioeconomic class children, born between the years 1970 and 1980 and followed longitudinally. The significant differences noted in both boys and girls between the older 50th percentile curve and the new curve indicate that the existing standards need to be revised. These

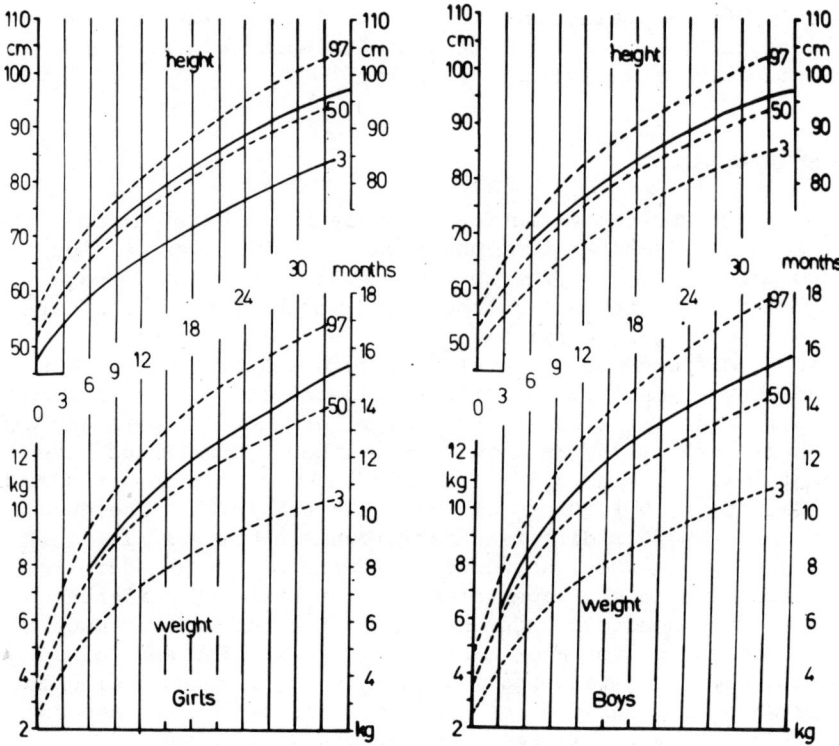

Figure 1. Secular trend in Turkish children.

data also indicate that the genetic potential for growth has not yet been reached in Turkish children.

Dr. Brandt has stated in her report that secular trend is very probably an outcome of improved environment in the first two years of life. Others also have pointed out that, due to the high growth rate that characterizes this period, a significant percentage of the biological and behavioral variability encountered in later years has its genesis in the preschool period of life (Heald and Hollander 1965, Huenemann 1974, Johnston 1978). In accordance with these considerations, protein energy malnutrition also produces its most serious and profound effects upon growth patterns during the first two to three years of life.

The outcome of severe undernutrition in animals has been shown to vary with the timing, duration and severity of the state of malnutrition (Widdowson 1963). In his studies on infants and children with celiac disease, Prader has shown that in states of nutritional deprivation, the organism is able to recover and catch up in growth when adequate nutrition is reinstituted during a developmental period of high growth potential (Prader 1978). Catch up growth has been reported in infants with marasmus and kwashiorkor by a number of authors. However there is no general agreement whether growth retardation caused by malnutrition is a reversible or irreversible phenomenon (Stock and Smythe 1963, Monckeberg 1968, Graham 1968, Yatkin and McLaren 1970, Briers 1974, Cravioto and Delicardie 1978, Branko 1979). Almost all investigators agree that intellectual development generally remains impaired in severe PEM. A group of 51 patients were evaluated after a mean period of 11 ± 2 years following discharge from our hospital with a diagnosis of kwashiorkor or marasmus. Almost all of the children in this group had continued to dwell in a suboptimal environment following their discharge from hospital and no child had been seen regularly by a medical team. 39 % of these patients were under 3rd percentile for weight and 49 % under 3rd percentile for height. Head circumference was smaller than -1 SD of the normal mean for age in 72.5 % of the group, being under -2 SD in 35.3 %. I.Q. evaluation by the Stanford-Binet and WISC tests in these cases yielded a mean value of 84.6 ± 14.3 (range: 63 to 107, unmeasurably low in 2 cases). It must be emphasized that similar to SGA infants, the physical and psychosocial environment provided for these patients after recovery from the severe state is a major factor and perhaps the most important one, in the ultimate prognosis of infantile severe PEM. Dr. Brandt has shown that catch up growth, including increase in head growth, occurs in a significant proportion of cases of fetal malnutrition when such an environment is provided.

Until a state of "health for all" is reached worldwide, the drastic effects of suboptimal medical environment in infants and children will continue to exist.

REFERENCES

Babson, S.G., 1970, Growth of low-birth-weight infants, J. Pediatr., 77:11.
Babson, S.G. and Phillips, D.S., 1970, Fetal growth : Liveborn birth weights for gestational age of white middle class infants, Pediatrics, 45:937.

Brandt, I., 1978, Growth dynamics of low-birth weight infants with emphasis on the perinatal period, In : F. Falkner and J.M. Tanner (eds) "Human Growth" Vol. 2, Bailliere-Tindall, London, p.

Branko Z., 1979, Height, weight and head circumference in survivors of marasmus and kwashiorkor, Amer J. Clin. Nutr., 32:1719.

Briers, P.J., Hootweg, J. and Stanfield, J.P., 1975, The long term effects of protein energy malnutrition in early childhood on bone age, cortical thickness and height, Acta Paediat. Scand, 64:853.

Campbell, S., 1974, Fetal growth, J. Clin. Obstet. Gynecol., 1:41.

Cravioto, J. and Delicardie, E.R., 1978, Nutrition, mental development and learning, In : F. Falkner (eds) "Human Growth" Vol 3, Bailliere-Tindall, London, p. 373.

Creasy, R.K., Resnik, R., 1981, Intrauterine fetal growth retardation In : A.Milunsky, E. Friedman, L. Gluck (eds) : "Advances in Perinatal Medicine" Vol 1, Plenum Medical Book Co., New York and London, p. 152.

Drillien, C.M., 1970, The small-for-date infant : Etiology and prognosis, Pediat. Clin. North Am., 17:9.

Erdogan, S., Neyzi, O., Günöz, H., 1982, Postnatal gestasyon yasinin üç ayri yöntemle saptanmasi Tip Fak Mecm, 45:489.

Eveleth, P.B., 1978, Population differences in growth: Environmental and genetic factors, In : F. Falkner, J.M. Tanner (eds) : "Human Growth" vol 3, Bailliere-Tindall, London, p. 373.

Fancourt, R., Campbell, S., Harvey, D., et al, 1976, Follow-up study of small-for-date babies, Br. Med. J., 1:1435.

Fitzhardinge, P.M., Steven, E.M., 1972, The small-for-date infant I. Later growth patterns, Pediatrics, 49:671.

Fitzhardinge, P.M., Steven, E.M., 1972, The small-for-date infant II. Neurological and intellectual sequelae, Pediatrics, 50:50.

Garn, S.M., Clark, D.C., 1975, Nutrition, growth, development and maturation : Findings from the Ten State Nutrition Survey of 1968-70, Pediatrics, 56:306.

Gatson, A., Crowther, D., Harvin, D. et al, 1972, Development of infants with intrauteringe growth retardation, Clin. Res., 20:280.

Goldstein, H., 1971, Factors influencing the height of seven-years-old-children. Results from the National Child Developmental Study, Hum. Biol., 43:92.

Graham, G.G., 1968, The later growth of malnourished infants: effects of age, severity and subsequent diet, In : R.A. McCance and E.M. Widdowson (eds) "Calorie Deficiences and Protein Deficiences", J. and A. Churchill Ltd, London, pp. 301.

Gruenwald, P., 1964, Infants of low birth weight among 5000 deliveries, Pediatrics, 34:157.

Gürson, C.T., Neyzi, O., 1966, "Istanbul'un Rami Gecekondu Bölgesinde Cocuk Sagligi Konusunda Arastirmalar", Istanbul.

Hagberg, G., Hagberg, H., Olow, I., 1976, The canging panorama of cerebral palsy in Sweden, 1954-70 III. The importance of foetal deprivation of supply, Acta Paediatr. Scand, 65:403.

Harrison, R.F., Roberts, A.P., 1977, A critical evaluation of tests used to assess gestational age, Br. J. Obst. Gynecol., 84:98.

Heald, F.P., Hollander, R.J., 1965, The relationship between obesity in adolescence and early growth, J. Pediatr., 67:35.

Hertz, R., Sokol, R.J., Knoke, J.D., Rosen, M.G., Chik, L., Hirsch, V.J., 1978, Clinical estimation of gestational age : Rules for avoiding preterm delivery, Am. J. Obstet. Gynecol., 131:395.

Holmes, G.E., Miller, H.C., Hassnein, K. et al, 1977, Postnatal somatic growth in infants with atypical fetal growth patterns, Am. J. Dis. Child, 131:1078.

Howard, J., Parmelee, A.H., Kopp, C.B., Litman, B., 1976, A neurological comparison of preterm and full-term infants at term conceptional age, J. Pediatr., 88:995.

Huenemann, R.L., 1974, Environmental factors associated with preschool obesity. I. Obesity in six month old children, J.Am. Diet. Assoc., 64:480.

Johnston, E.F., 1978, Somatic growth of the infant and preschool child, In : F. Falkner, J.M. Tanner (eds) : "Human Growth" Vol 2, Bailliere-Tindall, London, p. 91.

Lindgren, G., 1976, Height, weight and menarche in Swedish urban school children in relation to socio-economic and regional factors, Ann Hum. Biol., 3:501.

Low, J.A., Galbraith, R.S., Muir et al, 1978, Intrauterine growth retardation : A preliminary report of long-term morbidity, Am. J. Obstet. Gynecol., 130,534.

Mata, L., Urrutia, J.J., Mohs, E., 1981, Implicaciones de bajo peso al racer para la salud publica. Arch Lationum Nut(Suppl 1) 27:198, 1977 (Cited by Cassady G. : The small for date infant, In : GB Avery (ed) "Neonatology, Pathophusiology and Management of the Newborn", J.P. Lippincott Co, Philadelphia-Toronto,p. 262.

Metcoff, J., 1978, Association of fetal growth with maternal nutrition, In : F. Falkner, J.M. Tanner (eds) "Human Growth" Vol 1, Bailliere-Tindall, London, p. 415.

Monckeberg, S., 1968, Effect of early marasmic malnutrition on subsequent physical and psychological development, In : N.E. Scrimshaw, J.E. Gordon (eds) "Malnutrition, Learning and Behavior" MIT Press Cambridge-Massachusetts.

Neyzi, O., Yalçindag, A., Alp, H., 1973, Heights and weights of Turkish Children, Journal of Tropical Pediatrics and environmental Child Health, 19:5.

Neyzi, O., Alp, H. Yalçindag,A., Yakacikli, S., Orhon, A., 1975, Sexual maturation in Turkish boys, Ann Hum. Biol., 2:251.

Neyzi, O., Alp, H., Orhon, A., 1975, Sexual maturation of Turkish girls, Ann Hum. Biol., 2:42.

Neyzi, O., Binyildiz, P., Alp, H., 1979, Growth standards for Turkish children : Heights and weights, Courrier, 29:553.

Neyzi, O., Alp, H., Yalçindag, A., Konrapa, A., Ertem, B., Celenk, A., Akdemir, C., Açikkol, K., 1979, Kentsel ve kirsal iki ACS merkezi materyelinin PEM yönünden degerlendirilmesi, Doga III: 93.

Neyzi, O., Günöz, H., Can, G., Avgan, I., Cehenk, A., Ozsarfati, J., Kutlu, N., Erol, A., 1979, Istanbul bölgesinde dogan yenidoganlara iliskin bazi veriler, In : "Türk Cocugunun Saglik Sorunlari", Ozdem Kardesler Matbaasi, Istanbul.

Neyzi, O., Günöz, H., Celenk, A., Ozsarfati, J., Sait, R., Yenerer, N., Uzel, T., Saka, N., 1982 (in press), Türkiye'de yenidoganlarda bölgesel büyüme ve olgunlasma farkliliklari, Tip Fak Mecm. (Sullp.).

Olowe, S.A., 1981, Standards of intrauterine growth for an African population at sea level, J. Pediatr., 99:489.

Ounsted, M., Scott, A., 1981, Associations between maternal weight, height, weight-for-height, weight-gain and birth, In : J. Dobbing (ed) "Maternal Nutrition in Pregnancy, Eating for Two ?", Academic Press, London-New York-Toronto-Sydney-San Francisco, p. 113.

Özalp, I., Ciliv, G., Erdem, G., Durmus, Z. and Dogramaci, I., 1980, Small for dates babies in 1018 consecutive births in a general maternity hospital in Ankara, Bulletin of the International Pediatric Association Vol. 3, No 5, p. 9.

Pape, K.E., Fitzhardinge, P.M., 1981, Perinatal damage to the developing brain, In : A. Milunsky, E.A. Friedman, L. Gluck (eds) "Advances in Perinatal Medicine" Vol 2, Plenum Medical Book Co New York and London, p. 61.

Prader, A., 1978, Catch-up growth, In : D. Barltrop (ed) "Paediatrics and Growth", Fellowship of Postgraduate Medicine, London, p. 133.

Stewart, A.L., Turcan, D.M., Rawlings, G., et al, 1977, Prognosis for infants weighing 1000 g or less at birth, Arch. Dis. Child, 52:97.

Stock, M.B., Smythe, P.M., 1982, Does undernutrition during infancy inhibit brain growth and subsequent intellectual development ? Arch. Dis. Child, 38;546.

Svenningsen, N.W., Lindroth, M., Lindquist, B., 1982, Growth in relation to protein intake of low birth weight infants, Early Human Development, Elsevier Biomedical Press, 6:47.

Tanner, J.M., 1966, The secular trend toward earlier physical maturation, T. Soc. Gneesk, 44:524.

Thomson, A.M., Billewicz, W.Z. and Hytten, F.E., 1968, The assessment of fetal growth, J. Obstet. Gynecol. Br. Commonw., 75:903.

Usher, R.H., 1981, The special problems of the premature infant, In : GB Avery (ed) "Neonatology, Pathophysiology and Management of the Newborn", J.B. Lippincott Co, Philadelphia-Toronto, p. 257.

Van den Berg, B.J., Yeroshalmy, J., 1966, The relationship of the rate of intrauterine growth in infants of low birth weight to mortality, morbidity, congenital anomalies, J. Pediatr., 69:531.

Van Wieringen, J.C., 1978, Secular growth changes, In : F. Falkner, J.M. Tanner (eds) "Human Growth" Vol 2, Bailliere-Tindall, London, p. 445.

Varma, T.R., 1973, Prediction of delivery data by ultrasound cephalometry, J. Obstet. Gynecol. Br. Commonw., 80:316.

Walker, J., Small-for-dates, clinical aspects, Proc. R. Soc. Med., 60:8.

WHO, 1980, Division of Family Health : The incidence of low birth weight. A critical review of available information, World Health Statistics Quarterly, No 3.

Widdowson, E.M. and R.A. McCance, 1963, The effect of finite periods of undernutrition at different ages on the composition and subsequent development of the rat, Proc. Roy. Soc. Lond., 158:329.

Yatkin, U.S., McLaren, D.S., 1970, The behavioral development of infants recovering from severe malnutrition, J. Ment. Defic. Res., 14:25.

DISCUSSION REPORT

(Medical Environment)

Joan Smith

Geneviève Massé: Intra-uterine Growth and Public Health
    Reply by Ingeborg Brandt

With regard to the question of comparability of "intrauterine" growth curves I should like to emphasize that in the Denver curves the shift towards lower levels created by the effect of height (1500 m) doesn't occur before 38 postmenstrual weeks.

Dr. Massé raises the question of social class factors (S.E.S.) contributing to birth at different gestational ages. The importance of these factors were found to be negligible in the Bonn study, there being no real difference in S.E.S. between our study and the rest of the population.

One is also conscious of the shortcomings of deriving velocity curves for weight and supine length from foetuses born at different post-menstrual ages and applying them as standards for the normal population. However, there is wide agreement between various independent sources for these distance values of intrauterine curves and we feel there is some justification in using them since there are no other data available.

I would also disagree with Dr. Massé that social class in any way affects the growth of the fetus in its earliest stages of development (conception to 20 weeks).

With regard to "abnormal" patterns of growth in utero, the use of the phenomenon of "flattening" as a risk factor is problematic. It is still unresolved how long retardation may be tolerated by the fetus without permanent damage. There is obviously a need for more research in this area particularly with regard to the timing of clinical intervention.

With regard to the problem of length as a risk indicator, it should be stated that length is only a good risk indicator if used together with weight. This may be demonstrated by the weight/length ratio of Rohrer, $I = \dfrac{100 \times \text{weight in g}}{(\text{supine length in cm})^3}$. If the index is significantly lowered, this may be taken as an indication of risk.

The measurement of crown-rump length is indeed felt to be an extremely accurate means of assessing maturity in the first trimester of pregnancy, since growth retardation at this early stage is not found in normal fetuses. The only possible exceptions to this general rule are fetuses with abnormal chromosomal complements or early infections.

The question of the estimation of brain weight from the biparietal diameter (BPD) was raised. However, work by Dobbing and Sands (1978) who calculated brain weight from the BPD using data from Campbell (1976) has shown a reasonable relationship between these two parameters.

Dr. Massé raised the question of preventive care. It was stated that this was really the province of the obstetrician but the usual procedures included regular prenatal examinations, ultra-sound measurements, counselling of the mother, bed rest if required, and hormonal controls.

A good relationship has been demonstrated between lack of catch-up growth in head circumference and a below average mental performance. So, in answer to Dr. Massé's point, there does seem to be an association between brain quantitative normal growth and the development of mental capacities.

The question of the prognosis for children without catch-up of head circumference was raised. In cases where there was a stimulating environment, optimal care and nutrition, there does seem to have been a positive influence on the later development of SGA infants without catch-up of head circumference. However, in judging this outcome one has to consider the great deal of inter-individual variation which is exhibited in normal development and there remains much to be discovered in this area.

Campbell, S. (1976): Fetal growth. In: Fetal Physiology and Medicine, pp. 271-301. R.W. Beard & P.W. Nathanielz Eds.

Dobbing, J., Sands, J. (1978): Head circumference, biparietal diameter and brain growth in fetal and postnatal life. Early Human Development, 2, 81-87.

Luigi Greco, Maria Mayer-Greco: Growth and Catch-up in Chronic Diseases of Childhood
  Reply by Ingeborg Brandt

Question 1 : When comparing full term with AGA infants it appears a very interesting suggestion: secular trend may be mainly based on the environmental conditions in the first 2 years of life. Does it mean that the definition of the growth target of an individual is done permanently at the early ages? Do these early conditions modulate the expression of the individual genetic potential of growth?
Answer : Here I would like to refer to the report of Olcay Neyzi (page 8) where she agrees with my suggestion that "due to the high growth rate that characterizes this period, a significant percentage of the ... variability encountered in later years has its genesis in the preschool period". I would suppose that puberty with its high growth velocity offers another opportunity to the individual to change the original growth channel provided the environmental conditions change significantly.

Question 2 : AGA infants do catch up their growth before 2 months of age, it appears that velocity is inversely proportional to their deficit of growth : it would be interesting to know what is the dispersion around the 2 months : do some AGA catch up quite later?
Answer : Here one has to differentiate between the different growth parameters which show different patterns in AGA preterm infants: (a) head growth is not retarded; (b) weight growth shows an initial retardation until the age of one month, from two months there is no significant difference between AGA preterms and their full term controls ; (c) supine length is significantly less in the AGA preterms until 18 months, subsequently there is no significant difference between both groups.

Question 3 : The presence, or absence of catch up growth (length or cranial) is used as criterion to discriminate between good and bad prognosis in SGA infants: could it be possible that this is a general biologic mechanism present in most conditions affecting growth? If so, an accurate study of the catch up growth in most conditions should be favoured.
Answer : Here again a differentiation has to be made between catch-up growth in supine length/height and head circumference. Contrary to catch-up growth in head circumference, which seems to be possible only during the first three to six months - a period of high growth velocity - catch-up growth in height could be observed also at later ages and independently whether catch-up has or has not occured in head circumference during the first months of life.

Brandt, I. (1981): Brain growth, fetal malnutrition, and clinical consequences. J. Perinat. Med., 9, 3-26.

Question 4 : Timing of catch-up seems also to be crucial since nutritional condition may improve length growth and not cranial growth: is there any suggestion about a "critical time" for the head to grow?

Answer : I can underline the important question about the "critical time" for head circumference - and therewith of the brain - to catch-up the intrauterine retardation. Little is known about the limiting factors of catch-up of head circumference besides inappropriate nutritional conditions. One limiting factor is the rapidly decreasing growth velocity of head circumference, another might be inappropriate nutrition immediately after birth and by this a postnatal prolongation of the prenatal state of undernutrition - in the sense of the "doubly" deprived rat of Winick - and with the additional risk of hypoglycemia (Brandt, 1981).

Question 5 : Factors affecting catch up in SGA and AGA are important : we come to the nutritional question: early and high energy feeds appear to be essential, I presume that these exclude breast milk, which is not so "high energy", despite mothers produce different milk for premature than for full term babies. I could find no mention on human milk in the report.

Answer : In our SGA preterm infants of very low birthweight, i.e. $\leq$ 1500 g, breast milk was not available. So the infants were fed cow's milk, adapted to breast milk but with a higher content of protein, which meets the requirements of the rapidly growing infant. The standard formula given contains per 100 ml: fat 3.3 g, lactose 8.6 g, and protein 2.3 g with a casein/lactalbulmin quotient of 40/60, similar to breast milk.

Question 6 : I would appreciate some comments on these babies who do not catch up: are they different for nutrition, prenatal environment, postnatal care, diseases, heredity?

Answer : Those babies who did not catch-up in head circumference were mostly born before the introduction of early and high-energy feeding. At the same time there was also an amelioration of pre-, peri- and postnatal care, by this nutrition seems to be only one but an important factor. The hereditary and social background of both groups of SGA preterm infants shows no significant differences. The mean head circumference of the mothers in Group A of 54.5 cm (SD 1.2) agrees well with that of the mothers of Group B of 54,6 cm (SD 1.4). Between the fathers there is a nonsignificant difference of 0.65 cm. Therefore the growth differences between both groups cannot be ascribed to genetic differences.

The workshop on "Medical Environment" proved both informative and stimulating. Any direct questions raised by the participants on Dr. Brandt's presentation have been dealt with separately above. The purpose of this section is to summarise the remarks made by members of the audience and the ensuing discussion.

The comments made in general discussion could be classified into various categories. There was some mention of the problems of methodology, in particular the selection of bases for comparison of early growth curves (Masse) and also for the construction of reference curves. Professor Hoang in particular, advocated the instruction and use of "local" growth curves, relevant to the particular population under investigation. It was generally decided that although this is a very real problem and one which should always be considered in the collection of data, it is basically insurmountable. Dr. Brandt stated that, in fact, such variation as exists can be relatively easily recognised and controlled for.

Several speakers dealt with the general problem of catch-up growth. It was agreed that treatment, to be most effective, must start as soon as possible and was particularly important before the age of two years. The amount of catch-up growth, however, is recognised as being proportional to the severity of the initial deficit. Another problem aired was why some children catch-up and others do not (Greco), i.e. what are the actual physiological and genetic mechanisms involved. The general consensus was that this was an area which needed more research and that the answers had yet to be properly elucidated.

Professor Susanne and Dr. Ferro-Luzzi both pointed out that twin studies might be informative with regard to this question. Such evidence as exists appears to indicate that there are differences in catch-up growth with the two different sorts of twin pairs and monozygotic twins as a result of twin competition in utero may show differences at birth of up to 24 %, but, with appropriate care become very similar. Dizygotic twin pairs, however, are more discordant in their response to treatment and generally show as much variation as sib pairs. This seems to underline the importance of the genetic component in growth and again indicates the need for caution in extrapolating from "normal" growth curves.

The discussion then moved on to the use of growth models and catch-up growth as indicators of improved medical status. Dr. Greco put forward his models of coeliac disease and cystic fibrosis indicating the good association between progress of these diseases, and, subsequently, treatment of them, with standardised growth curves. Onset of each disease produces retardation in height and weight, while successful treatment shows catch-up in both these parameters, with weight being the more susceptible. Obviously, the amount of retardation and catch-up are proportional to the severity of the disease. Professor Tanner indicated the similarity of Dr. Greco's model to his experiences with growth hormone deficient children. Professor Tanner and Dr. Greco agreed that reference to standardised growth curves could prove invaluable in assessing the progress of such medical conditions.

They should always be used, however, in conjunction with information on the patients' parents height i.e. with reference to the genetic component of attained height.

Professor Van Wieringen raised the question of secular trend and the problems of distinguishing between this phenomena and catch-up growth. He emphasised that secular trend may occur any time in the growth period when inhibiting factors are not in evidence, though once again, Dr. Brandt stressed the importance of the first two years of life. Professor Tanner pointed out the relatively low correlation (0.7) in well-nourished normals between growth pre- and post-puberty. This, he said, makes the prediction of adult height problematical because it is as yet not possible to allow for the presumably inherited variable contributing to the adolescent growth spurt. It is not unusual for children to move from one growth channel to another pre- and post-puberty for no clinical reason. Indeed, it was stated by Professor Tanner, there is still confusion in the literature between secular trend and catch-up growth around the adolescent growth spurt. For example, the Harpenden Growth Survey found no evidence of catch-up at adolescence for Indian children living in Britain, a finding which was thought to indicate possible racial differences in growth patterns. However, data on Australian aborigines (e.g. Brown, 1978) shows a definite adolescent spurt in tooth eruption and Roche also reported verbally a tremendous recent secular trend in other parameters in Australian aboriginees, presumably, he believed, due to improvements in health care and nutrition. Thus the recent data on aboriginees would seem to indicate that perhaps these observed racial differences are in fact more to do with nutritional and environmental factors around puberty such as hard physical work, precocious sexual intercourse and inappropriate nutrition, than any inherent variation in growth patterns. This is, of course, not to imply that given optimum conditions, that all populations would attain similar mean height, but rather that patterns of growth might be similar.

Several participants asked what were good indicators of growth, particularly with regard to the study of catch-up growth. Dr. Brandt emphasised the use of head circumference, which due to the interlocking of sutures by about six years of age, can only be investigated in early childhood. Other speakers mentioned height and weight, though the latter did appear to exhibit much more variability.

Professor Neyzi outlined the situation in the country (Turkey) where the medical environment, in contrast to that described by Dr. Brandt, was not optimum. Her studies indicated that growth suffers most when growth velocity is at its highest i.e. ante- and post-natally and puberty. A poor medical environment and, in particular, protein energy malnutrition in Turkey produced approximately 30 % of term SGA infants with a handicap and another 30 % with growth stunting. She emphasised that growth curves vary in underdeveloped countries with higher social class exhibiting growth patterns more similar to Western European and North American findings. However, surprisingly, improved medical care produces secular trend in all social classes, though obviously more spectacularly in the lower groups.

The session concluded with a number of participants stressing the need for closer attention by medical practitioners and health care officials to some of the issues raised and more use of the recognised parameters to isolate those individuals beginning to suffer growth retardation for any reason and, subsequently, to monitor and assess their progress.

Brown, T. (1978): Tooth emergence in Australian aboriginals. Ann. Hum. Biol. 5, p. 41-54.

Ecological Environment

ENVIRONMENT AND PHYSICAL GROWTH

Anna Ferro-Luzzi

Unit of Human Nutrition
National Institute of Nutrition
Via Ardeatina, 546 - Rome, Italy

This paper is intended to be a critical review of the recent literature on environmental causes of growth retardation in man. It does not intend to be an exhaustive account of all published material nor of all the known causes of growth retardation. A selection has been made of the issues most relevant to public health and to population groups, neglecting the 'rare' aspects of more academic than practical interest.

Adverse environmental conditions can interfere with the full expression of the genetic programming for size of the individual and cause growth failure. The human phenotype represents the cumulative evidence of the environmental stress load experienced throughout the evolutive period. Under stressful conditions, small body size is considered to represent the outcome of an adaptive strategy. The cost/benefit ratio of this survival strategy may be questioned. Some authors have considered stunting as beneficial in terms of populations (Monod Cassidy, 1980; Abbott Seagraves, 1977; Frisancho et al., 1973). This approach tends to isolate the phenomenon from the context and to neglect the fact that smaller body size is the response to adverse environmental conditions. Under these conditions, smaller body size in children has been found to associate with permanent impairment of mental development, reduced work capacity and higher morbidity and mortality rates. A more comprehensive approach, scientifically as well as ethically correct, is the one that regards stunting as a waste of human resources and as an indicator of undesirable environmental pressures.

The environmental stressors that have been associated with growth failure are: nutritional factors, infectious diseases, high altitude hypoxia, socioeconomic conditions, smoking during pregnancy, neurohumoral disturbances secondary to psychological deprivation, temperature or climate extremes.

A nearly infinite combination of duration, severity, timing and nature of these stressors is theoretically possible. Each combination is bound to stimulate in a different manner the adaptive processes of the organism, eliciting a wide array of different responses.

The timing of the exposure to stress plays an important role in the nature and degree of the response of the organism. Growth is an uneven process whose rate is regulated by a maturational "clock" (Tanner, 1963). Disturbing factors may interfere, either by delaying the maturation process (timing effect) or by directly preventing the fullfilment of the potential for growth (intensity effect). Maturation may be delayed to a lesser extent than growth, with different consequences on body size. The increasing inhability of the growing animal to achieve full genetic size with nutritional rehabilitation following increasingly longer periods of undernutrition has been attributed to the timing effect (McCance and Widdowson, 1974). In man, linear growth retardation was found to be more delayed than skeletal maturation in chronically malnourished Ladino children (Martorell et al., 1979). Food supplementation improved both growth performance and skeletal maturity; however, the effects on body size (intensity of growth) were more marked than those on skeletal maturity (timing effect). These findings are consistent with the results obtained by Frisancho et al.(1970) in rural populations of Central America; growth retardation was found to be most marked at ages 10-15; retardation in skeletal maturation, on the other hand, ranged from 12 to 39% during early and later childhood, but only 5 to 9% during adolescence. The potential for continued growth was thus less retarded than the actual growth, limiting accordingly the extent of catch-up growth. The growth period was calculated to be prolonged by about 7% as compared to the standards.

INDICATORS OF GROWTH PERFORMANCE

Weight and height are the conventional indicators of body size and have been universally employed as dependent variables in the predictive equations describing the impact of environment on genotype. Both have been used as such or as part of composite indices. Weight and, especially, height provide simple cumulative measures of growth. Other indices have been occasionally used: skinfold thickness as proxy of body fat; upper arm muscle area or circumference as proxy indicator of protein reserves; sitting height and laterality indices as indicators of body proportions.

Birth weight, corrected for gestational age, has been used as indicator of antenatal growth. Recently, new techniques for intrauterine monitoring of foetal development processes have been developed (Campbell, 1974), but are still too avant-garde to be widely used.

DETERMINANTS OF GROWTH PERFORMANCE

The interest of assessing the discrete impact of the main contributors of a multistressor environment upon physical growth derives from the need of providing concrete guidance for intervention policies as well as from an academic concern.

For practical purposes, the environmental stressors may be grouped in two major categories: direct and indirect determinants. The direct determinants are food intake and infectious diseases. Both disturb the nutritional status and act through metabolic perturbations that adversely effect nutrient utilization: nutrient sequestration or wastage, increased energy and/or nutrient requirement, losses from the body, mobilization of stores, etc. The indirect determinants are all those conditions that affect food intake and that facilitate the spread of infections, such as level of income, water supply, waste disposal, overcrowding, family size, belief patterns relating to feeding practices, etc. A separate and special category of indirect determinants is represented by high altitude hypoxia and smoking during pregnancy.

Direct Determinants

1.Infections. The intimate association between growth and infection has been fairly established by animal models as well as by field studies on human populations. However, the precise nature of the relationship remains to be defined. The multitude of interacting environmental factors in population studies, the variability of severity and duration of diseases, the nature of the microbes, all concur to complicate the picture. The attempts to establish a direct causative link between growth failure and infection have often produced conflicting results. Further complications derive from the fact that the socioeconomic and cultural background, characterized by low income, polluted water, lack of waste disposal systems, etc. that favours the uncontrolled spread of infections, is the same that breeds high prevalence of malnutrition. This common matrix explains the high degree of covariation of socioeconomic indicators, growth and morbidity. Nutritional deficiencies, by impairing several immune responses may facilitate, in turn, the insurgence of infectious diseases (Chandra, 1979).

A list of the nature and timing of the metabolic-endocrine responses of the host to infectious processes has been drawn by Beisel (1975) who distinguishes catabolic losses, or absolute wastage, from functional or internal wastage of nutrients. Diarrhoea, vomit, loss of appetite and haemorrhages have an additional effect. In the ultimate analysis, the mechanism by which infectious diseases interfere with growth processes appears to be *via* nutritional disturbances (Fig. 1).

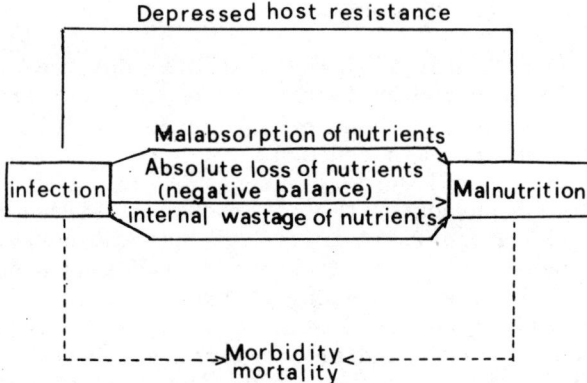

Fig. 1. Schematic representation of the reciprocal interrelationship between infection and malnutrition, with emphasis on infection-related causes of malnutrition (reproduced with permission from Beisel, 1975).

Infections experienced during intrauterine life can cause foetal growth retardation, besides other more serious and even terminal effects, such as interruption of pregnancy, prematurity, embryopathies of various kinds (Katz et al., 1975; Alford et al., 1969; Mata et al., 1972; Mata et al., 1977). Elevated cord Immunoglobulin M (IgM) levels are an index of fetal response to antigenic stimulation (Alford et al., 1969; Chandra and Matsumara, 1979; Lechtig and Mata, 1971). High values of cord blood IgM have been found in the cord blood of the newborn in developing countries (Mata et al., 1977; Logie et al., 1973) in association with high prevalences of birth weights less than 2500 g (Mata et al., 1977) suggesting an important role of prenatal exposure to infectious stresses in terms of impairment of physical growth. It would be difficult, however to separate the effects of intrauterine infections from those related to maternal malnutrition.

A vast body of evidence exists in the literature on the association between postnatal illness history and growth failure. Martorell (1980) has recently reviewed the epidemiological evidence from field studies and found that, while in developed countries the association between illness and physical growth tends to be inconsistent or absent, in developing nations common childhood illnesses, particularly diarrhoeal diseases, are usually associated with growth retardation. Diarrhoeal infections proved to be statistically significant determinants of growth failure in Indian children 6 to 24 months old (Levinson, 1977). In rural South America the occurrence of infections was found to precipitate or deteriorate the nutritional status of the affected child and to produce weight losses and growth arrest (Mata et al., 1977). The contribution of the effect of diarrhoeal diseases to growth retardation was estimated, in a prospective study on Ladino children, to range around 10% (Mata et al., 1972). The nature of the infectious diseases was found to elicit differential responses in the growth processes of the host, e.g. viral diseases affected weight earlier and more rapidly than height (Mata et al., 1972); *Giardia* infestation proved to be non significantly related to growth failure in Indian children (Levinson, 1977). Retardation in linear growth of about 2.2 cm, equal to about 4 months of growth, was found to be associated in Guatemalan children with radiopaque transverse lines at the distal end of the radial metaphysis (Blanco et al., 1974). Lines of dense calcium deposition are considered to be indicative of disturbances in the normal chondroplastic processes and to indicate growth arrest (Acheson, 1959; Hewitt et al., 1955). Originally observed by Harris (1931), these lines were produced in the long bones of experimental animals in response to a variety of acute stresses of various nature such as severe infections, protein-energy malnutrition, minor surgery, etc.

2.Malnutrition. Nutrition occupies a position of prime importance in the composite multistress environment of man, being a primary stressor agent in its own right as well as a powerful mediator or functional correlate, at the cellular and biochemical level, of the adaptive responses of the organism to various other environmental stressors.

It is well documented that chronic marginal malnutrition is the most frequent form of environmental stress and that wide sectors of humanity, subsisting in conditions of poverty, are exposed to it. In these populations growth performance appears to be seriously disturbed, lagging behind that of well-fed populations and the adult body size is usually smaller.

Complex relationships have been revealed to exist between stunted growth and deficits in cognitive and in work performance morbidity and mortality. Different risks were found to be associated with different degrees of stunting. Body size has been thus used as proxy indicator for nutritional status, and the prevalence of protein-energy malnutrition in a community is currently assessed by means of anthropometric indicators (Keller et al., 1976). The diagnostic value of these indicators has been abundantly validated by animal experimentation and by controlled nutritional rehabilitation studies in children.

A wide range of variation exists in the size of the individuals within any population. In well-fed healthy populations parental and child dimensions show the high degree of intrafamilial correlations that is expected when the genetic component dominates the phenotypic expression (Beargie et al., 1970; Tanner and Israelsohn, 1963; Livson et al., 1962; Garn and Rohmann, 1966; Wolanski, 1970; Mueller, 1976; Welon and Bielicki, 1971). Being the expression of genetic programming, small size in these individuals does not appear to have any relation to negative functional or cognitive parameters.

In populations where body size is reduced in response to environmental stressors, the wide range of variations observed in the relationship between growth failure and nutrition may reflect different individual adaptive capacities or different timing of exposure; the variability may also derive from the interplay of the nutritional stressor with a variety of other adverse factors capable of intensifying or precipitating the response. Finally a critical level of stress may be needed before the child genetic programming becomes dominated by environmental forces.

Before moving on to describe the nature of the relationships between nutritional environment and growth failure, it is appropriate to mention the possibility that part of the observed deviations of body size from the norm may be the genetic expression of ethnic differences. At present, the issue remains still somewhat controversial. Evidence is available showing that the growth performance of healthy children from economically privileged sectors of different ethnic groups may be comparable to that of Western reference standards (Johnston et al., 1973; Habicht et al., 1974; Aschcroft et al., 1966). Linear growth of healthy well-fed Guatemalan children has been proved to be similar to that of socioeconomically matched peers of European extraction (Johnston et al., 1977). Growth patterns however became significantly different during the adolescent period, suggesting a late emergence of a genetic component. Similar findings were reported in the Peruvian lowlands, when 6 to 19 year old Mestizos and Quechuas were compared (Frisancho et al., 1980).

In conclusion, although a case can be made for either hypothesis, the available evidence suggests that ethnic growth potentials, where they exist, do in fact play a secondary role, at least up to adolescence, as compared to the overriding influence of environmental and nutritional stresses. Thus, the departure of physical dimensions from the norm may be safely accepted as a measure of adverse environmental conditions.

The scientific purity of field observations on growth and nutrition inevitably suffers from the complexity of the multistressor environment, as revealed by the high variability in the strength of the observed correlations. A note of caution should be here raised for what concerns the use and interpretation of dietary consumption data. It must be understood that nutrient intake data identify nutritionally at risk subjects, i.e. those whose energy and nutrient requirements are not met. Consumption data per se do not provide any indication about actual nutritional status. This great limitation of the diagnostic value of dietary intake data may help in understanding the sometimes observed lack of relationship between attained size as dependent variable, and energy and nutrient intake data. Also, while attained size reflects the cumulative life-long exposure to a multiplicity of stresses, the dietary intake data are usually assessed over limited, and usually brief, periods of the life span.

Table 1. Energy and nutrient adequacy and anthropometric indicators of a group of 1-6 year old Italian children classified either according their nutritional status (based upon growth criteria) or as being nutritionally at risk (on the basis of their dietary intakes) (from Ferro-Luzzi et al., 1979).

	WELLNOURISHED	MALNOURISHED	AT RISK
Number of Children	434	77	208
ADEQUACY (%) OF INTAKES OF:			
Energy	93	92	62
Protein	154	150	108
Calcium	126	130	93
Iron	84	81	58
ANTHROPOMETRIC INDICATORS :			
Weight for Height %	102	87	100
Triceps for Age %	108	76	95
Arm Muscle Area cm^2	17	14	16

Extrapolation of the findings to a remote past relies on indirect evidence and may prove to be inappropriate (Ferro-Luzzi, 1982). Examples of lack of association between growth performance and dietary intakes may be found in the literature. The dietary intakes of stunted Italian children of preschool age shown in table 1 were similar to these of children with normal body size; conversely, children whose diet classified them as being nutritionally at risk, exhibited growth performance that were normal and comparable to those of children with satisfactory dietary intakes (Ferro-Luzzi et al., 1979). No difference could be detected in the nutritional adequacy of the dietary intakes of another group of Italian children whose growth performance was shown to reflect their contrasting socioeconomic background (Ferro-Luzzi and Mariani, 1981). The dietary intakes of 1 to 6 year old Black and White North American children were found to be strongly correlated to socioeconomic class, the lowest classes consuming the poorest diets (Owen et al., 1974); as the greatest majority of Black children belonged to the two lowest socioeconomic classes they may be assumed to subsist on the poorest diets. This assumption conflicts with the better growth performance of the Black children, who were taller, heavier, and incidentally leaner, than the White children.

2.1. *Birth weight and maternal nutrition.* Adequate availability of nutrients and energy to the pregnant woman is probably the major single prerequisite for optimal foetal growth (Lechtig et al., 1975). It has been assumed that when food intake is insufficient, the biological distribution of nutrients between the maternal and foetal organism might be tilted in favour of the foetus. The assumption is based on the hypothesis that organs and tissues with higher metabolic rates drawn proportionally larger shares of a limited nutrient supply (Naismith, 1969). That this may not be necessarily the case has been suggested by a series of experimental studies. The maternal organism was found to be able to succesfully compete with the foetus in presence of restricted food availability and protect herself from the excessive transfer of nutrients and consequent depletion, thus contraddicting the current concept of active foetal parasitism (Rosso, 1981). In the pregnant rat special mechanisms would be activated by which the rate of delivery of nutrients to the foetus is reduced (Rosso, 1977a; Rosso, 1977b; Rosso, 1981; Lederman and Rosso, 1980). This mechanism appears to obey to somewhat different regulators, according the severity and the nature (wether protein, energy or protein-energy) of the deprivation. The existence of a critical cut-off point has been postulated, below which the mother is favoured over the conceptus in this biological competition (Rosso, 1981). Human data collected during the Dutch famine were reexamined and found to be consistent with the animal model results (Rosso, 1981). Besides animal studies and retrospective epidemiological evidence in man, there are in the literature several prospective studies that have attempted to relate controlled dietary intervention on large samples of population exposed to nutritional pressure, to pregnancy outcome (McDonald et al., 1981; Morà et al., 1979; Rush et al., 1980; Lechtig et al., 1978). The results of these studies are somewhat ambiguous. No clear-cut support was obtained for the study hypothesis that increased energy or protein intake during pregnancy does positively influence pregnancy outcome. Only in the case of the Guatemalan study (Lechtig et al., 1978), the negative correlation between energy supplementation and incidence of low birth weight infants was statistically significant, although some degree of increment in birth weight was observed in all studies. In the Guatemalan study, the Authors report an increment of 3 g in the average birth weight of the infants for each 1000 kcal of supplement to total energy intakes during pregnancy (Lechtig et al., 1978). It is important however to remember that only less than 1/3 of the observed total variance of birth weight could be explained by environmental and maternal parameters; maternal dietary supplementation accounted for no more than 12% of the explained variance in birth weight (Habicht et al., 1974).

Other epidemiological non-intervention studies have relied upon conventional indicators of nutritional status in order to explore the effect of maternal nutrition on foetal development. Upper arm muscle area was taken as proxy for the level of body protein reserves and maternal subcutaneous fat was used as proxy of her energy stores. A statistically significant relationship was observed by Frisancho et al. (1974) in a large sample of Peruvian urban population between birth weight and protein and energy reserves of the mother. The greater the maternal upper arm muscle area the greater the linear dimensions of the offspring while maternal skinfolds had more influence on the fatness of the infant. In yet another study (Metcoff, 1981), about 75% of the variance in birth weight could be accounted for by stepwise multiple regressions, when 25 variables describing maternal nutritional status at mid-pregnancy were incorporated, and birth weight was adjusted for non-nutritional conditioning factors (such as gestational age, sex, parity, maternal age, height and pre-pregnancy weight).

2.2. *The role of specific nutrients.* The role of specific nutrients in growth performance has been extensively investigated in the experimental animal and has produced a vast body of literature. In human populations, however, the picture is much less clear. Besides energy, protein, iodine, vitamin D, and, more recently, zinc, no other nutrient has been firmly established as causative of growth retardation. Iodine deficiency, severe enough to cause hypothyroidism, has been implicated in growth retardation, acting at the level of the growth cartilage. Vitamin D deficiency (rickets) can cause stunting. Neither of these two pathologies will be treated in this review.

Zinc is an essential nutrient (Underwood, 1971) entering in the composition of numerous metalloenzymes. Its activity is related to nearly all aspects of metabolism, the synthesis of protein and nucleic acids, the mobilization of vitamin A from the liver. It is well established that dietary zinc deficiency prevents normal growth in experimental animals, and it has been associated with growth retardation and failure of sexual maturation in adolescent boys. The role played by zinc in the process of growth is not clearly understood. Primary dietary zinc deficiency was first reported in Egypt (Prasad et al., 1963). Treatment of zinc deficient Iranian subjects produced dramatic increases in linear growth (Halsted et al., 1972). Middle and upper income North American children with poor growth performance were found to be zinc deficient (Hambidge et al., 1972). Supplementation with zinc promoted growth in infants with low plasma zinc levels (Walravens and Hambidge, 1976). Growth faltering was found to be associated in low income Jugoslavian (Butzina et al., 1980). and Denver (Hambidge et al., 1976) children with lower zinc values in hair and plasma.

Being the commonest form of malnutrition, energy and protein deficits occupy by far the most important position. The respective role of protein and energy deficits on human growth has been investigated but is still object of controversy. The contention that retardation in linear growth is caused by protein deficiency (Graham et al., 1981; Campbell, 1974) does not seem to be supported by the dietary evidence. Protein or energy supplementation studies in growth retarded children has produced conflicting results, the rates of growth being responsive in some groups to energy supplementation (Gopalan et al., 1973; Martorell et al., 1978) in others to protein supplementation (Malcolm, 1970). Energy intake correlated better than protein intake with growth performance of 6-24 months Indian children (Levinson, 1977). Graham et al. (1981) derived multiple

regression equations in a group of poor urban children, where the
dependent variable was growth outcome (expressed as a variety of combinations of height, and weight indices) and the independent variable
was food intake. The regression analysis indicated a strong effect of
the percentage of dietary animal protein on linear growth, in spite
of the finding that protein intakes of most subjects exceeded the recommendations. The interpretation given to this unexpected finding
was entirely speculative and postulated, besides a specific effect of
dietary protein on linear growth, the association of protein in foods
to nutrients with growth promoting effects, or a digestibility factor,
or finally the underestimation of current protein intake recommendations.

The apparently contradictory findings on the respective role of
energy and protein in causing growth retardation or in promoting
catch-up growth may find an explanation in the different dietary backgrounds of the investigated groups and/or in the presence in the
environment of not-accountedfor confounding variables.

Indirect Determinants

1. <u>Socioeconomic factors</u>. The socioeconomic environment is the
dynamic expression of the complicated and multidirectional interplay
of elements of the social organization. Abundant literature exists
on the impact of socioeconomic conditions on growth (Valverde et al.,
1981; Levinson, 1977; Johnston et al., 1980; Ariza Macias et al.,
1973; Ferro-Luzzi, 1967; Greiner and Latham, 1981; Bogin and McVean,
1981; Lowstein and O'Connell, 1974). However, sophisticated analytical approaches are needed to identify and disaggregate the individual
effect of the main components of the environment and to measure their
independent contribution. As no single environmental variable can independently account for more than a minimal proportion of the variance
in growth performance, most analytical models include a wide
number of socioeconomic, cultural and demographic indicators. A set
of 40 socioeconomic, demographic, anthropometric and morbidity variables have been used by Johnston et al. (1980). Those variables that
loaded highest in a principal component analysis were regrouped in
5 factors. The proportion of variance of growth parameters that could
be accounted for by the direct or joint effect of the selected
factors was then calculated by a series of path analyses. The socioeconomic factor correlated significantly and showed an increasing
contributory effect with advancing age, suggesting a cumulative
effect in postnatal life. The variables included in this factor were:
food expenditure, in absolute as well as in proportion of income, parental education and hygiene, and a sanitation score. Of the other 4
factors, maternal reproductive history and illness failed to show any
significant effect. Parental sizes were significantly related and a
discriminant function analysis proved that the laterality index (biacromial and bicristal breadths) contributed most heavily to discrimi-

nate between children with and without growth failure. The variance in the laterality index was considered to be suggestive of parental exposure to nutrition deprivation during the developmental stages and their discriminant power was assumed to mean that their children were experiencing the same nutritionally adverse environment.

Per capita income was found to be better related to energy intakes of growth retarded 6-24 months old Indian children, than a more precise wealth index, suggesting that income liquidity may be more important than accumulated assets (Levinson, 1977). Per capita income was also found to have an independent effect on growth, above and beyond its effect on energy intake and infection index. The "living level", a composite index obtained by summation of a number of variables related mostly to housing conditions, proved to have the greatest impact on body size (weight for age) of Eastern Caribbean children; next in importance came the age of weaning and the number of siblings (Greiner and Latham, 1981).

Twentyfour to 28% of the variation in weight and in height of 1 year old Jamaican children was accounted for by a family demographic stress factor (number of siblings and household size), the degree of dependence of the household on subsistence agriculture, a maternal-gardian maturity factor and other 5 environmental factors (Marchione and Prior, 1980). These same factors loaded quite differently and predicted only 11 to 17% of the variance when the survey was repeated, with the same technique and in the same towns, two years later, after profound "if not revolutionary social and cultural change" had occurred in the Jamaican society.

Socioeconomic differences in body dimensions have been reported also in several developed countries. Seven year old British children of higher social classes were found to be 3.3 cm taller than those from the lowest (Goldstein, 1971). When several covariants were controlled (birth weight, mother's stature, age, parity, smoking during pregnacy, gestation, and number of siblings) the adjusted difference in stature was still 1.3 cm. No socioeconomic difference in weight or height was found in urban Swedish children; however geographical differences in growth were found whose meaning and cause remained matter of speculation (Lindgren, 1976). The analysis of growth data relative to 40,000 children proved that North American lower class children still tend to be smaller and lighter than children of higher socioeconomic background (Garn et al., 1978). The dimensional differences were already present at birth and increased progressively through age 7 indicating a cumulative environmental effect. White and Black children were analysed separately since they differed in size. These findings are consistent with the socioeconomic differences in stature and weight reported by Schutte (1980) for 10-18 year old Black American males.

The dynamics at the basis of these socioeconomic differences in

size is not clearly understood, as illustrated by the high number
and variety of indipendent indicators used in the predictive equations. The potential spuriousness of the correlations, the possibility that some of the independent variables may be, in fact, proxy
expression of others, and the inconsistency in the predictive power
of socioeconomic factors under different ecological settings are
cause of some concern. One note of caution to this effect comes from
Marchione and Prior (1980) who warn that "social ecological predictors can be quite misleading, unless the diachronic dynamics of the
society under study are understood".

2. <u>Cigarette smoking during pregnancy</u>. Tobacco smoking during
pregnancy can cause intrauterine growth retardation (Meyer et al.,
1976). A comprehensive paper on the influence of maternal smoking on
birth weight and length has systematically reviewed the results of
about 50 studies covering a total of nearly 140,000 infants (Meredith,
1975). Only industrial countries are included in the paper as a negligible minority of women in developing countries do smoke. Offspring
of women who used tobacco during pregnancy were shown to weigh at
birth 170 g less and be 0.4 cm shorter than infants born to non smoking women. The effect of tobacco on foetal development was shown to
be graded according the number of cigarettes, the children born to
mothers smoking 10 or more cigarettes weighing 200 g less than children born to non smoking mothers. Smoking 20 or more cigarettes produced a difference of 250 g. The consequence of this downward shift
of birth weight involved, obviously, an increase in the number of
children weighing 2500 g or less. While only 5.7% of the progeny of
non smoking mothers did fall in that category, the percentage for
smoking mothers increased almost two fold. The statistics, however,
did not control or discriminate for socioeconomic status of the mothers or for sex of the child; it is plausible that some confounding
variable may have interacted to some extent with the effect of smoking on foetal development.

The effect of maternal smoking on the size of the progeny has
been found to be long lasting and to act above and beyond its effect
on birth weight. British children born to mothers who did not smoke
during pregnancy, were 0.6 cm taller at age 7, after controlling for
various confounding variables, than those born to heavy smokers
(above 10 cigarettes) (Goldstein, 1971). After allowing for the difference in height that could be accounted for by the lower birth
weight, smoking during pregnancy appeared to cause a decrease in
stature of 1.0 cm.

3. <u>High altitude hypoxia</u>. Several research directed to elucidate
the role played by hypobaric hypoxia on foetal development, conducted
mostly in the South American Andes, have provided good evidence that
women with life-long exposure to high altitude give birth to infants

that are significantly lighter and shorter than low-altitude infants
(Haas et al., 1977; McClung, 1969; Haas, 1980; Haas et al., 1980;
Howard et al., 1957; Lichty et al., 1957). These newborns have also
reduced skinfold thickness (McClung, 1969), modified body composition
and retarded skeletal development (Haas et al., 1977), suggesting that
the high altitude effect is particularly marked during the last trimester of gestation (Haas et al., 1977). The dimensional differences
have been found to persist after the confounding effects of ethnic
group and socioeconomic conditions were controlled. The reduction in
birth weight ranged between 200 and 300 grams for altitudes exceeding
3000 m. The exact cause of this phenomenon in not completely understood. Exposure to high altitude hypoxia results in an immediate adaptive response of the oxygen-trasport system; if exposure is prolonged,
long-term adjustments of other functions are known to occur (lung
diffusion capacity, tissue vascularization, oxygen-haemoglobin binding
affinity, cellular metabolism (Mazess, 1975). Even at sea level the
foetus develops in a physiologically hypoxic environment (Eastman,
1930; Bancroft, 1930). Exposure to hypobaric hypoxia further reduces
the margin of safety of the oxygen supply to the foetus. During pregnancy at high altitude adaptive placental modifications aimed at
improving foetal oxygenation have been described (McClung, 1969;
Chabes et al., 1968). These adaptive processes are reflected in modifications of the placental shape, weight and composition. It has been
postulated that the multigenerational exposure to high altitude could
improve the adaptation, possibly by means of population selective
strategy (Haas et al., 1980). Haas et al. (1980) studied the influence
of hypoxia on birth weight of children born to Indian and non Indian
women at high and at low altitudes. When all relevant covariates were
controlled (parity, gestational age, mother's stature and arm muscle
area) a three way analysis of variance detected a relationship between
weight, altitude and ethnic group. Low altitude infants had, as
expected, mean weights that, after adjustment for the other dependent
variables, were 282 g heavier than high altitude ones; however Indian
infants were 143 g heavier than non Indian after controlling for the
other dependent variables, suggesting a superiority of the adjustment
achieved by the native Indian women as a consequence of millenia of
exposure to high altitude. A noticeable interaction between sex and
the altitude effect was also reported, males presenting a greater
altitude effect. The greater susceptibility of the male infants to
environmental stressor is consistent with findings of other authors
(Stini, 1972; Stini, 1981; Haas, 1974; Acheson and Fowler, 1964) and
with the theory of a greater stability in development and a more
canalized growth of the females.

FACTORS ASSOCIATED WITH GROWTH PERFORMANCE IN ITALY

The picture offered by Italy in the '40s was one where socioeconomic, dietary and infectious factors could be assumed to create remarkable pressure on the genetic potential for growth of the population. The GNP (Gross National Product) in 1951 was estimated to be 242,000 liras *per caput per year* (ISTAT, 1981 revised figures). The 1951-1960 national Food Balance Sheets indicated that the gross energy and protein availability amounted to 2620 kcal *per caput and per day* (alcohol excluded) and to 63 g *per caput and per day* respectively (Ferro-Luzzi and Sofia, 1962). This was estimated to be marginal as compared to theoretical requirements (Ferro-Luzzi, 1966). Clinical signs of classical deficiency syndromes (dystrophic oedema, rickets, pellagra, goiter, mehlnahrschaden, nutritional dwarfism) were reported as late as the '50s in various areas of the country, with higher prevalence in Southern and Insular Italy and in rural areas (Ferro-Luzzi and Sofia, 1962; Frontali, 1950; Bacchetta, 1951; Ferro-Luzzi, 1962; Mancini and Fiorentini, 1957). Against this adverse background, evidence of the expected growth failure was provided by several local studies as well as by national surveys. In those times the statistical procedures employed for analyzing epidemiological studies of growth performance were rather simple as compared with the sophisticated statistical methods now currently used, and the computer technology was not yet available. There can be, however, little doubt that the Italian child was then exhibiting evidence of marked growth failure and presumably not fulfilling its genetical potential.

At local level, a well documented example of growth failure in an adverse environment can be found in the Rofrano study (Cresta et al., 1982). Conducted in the '50s, it described the degree of adversity of the general living conditions in an area selected as typical of the economic and social backwardness prevailing in deprived rural areas of the South. Over 50% of the families were holders of less than 5 acres of land. Primitive agricultural practices were carried out on poor unfertile mountainous soil, producing wheat, olives and figs for family subsistence. Ninetyfour percent of the houses had no running water and 80% had no hygienic facilities. Thirtyseven percent of the population aged more than 6 years was illiterate. Intestinal parasitism (*Trichiuris trichiura, Giardia intestinales, Ascaris lumbricoides, Enterobius vermicularis,* etc.) affected 96% of the 6-12 year olds, and more than one child out of ten was affected by polyparasitism (up to 6 parasites). Twentysix percent of children had X-ray signs of rickets. Against this background, growth failure of the children was marked (Cresta and Fiorentini, 1956): the stature lagged 1.5 to 2 years behind the Italian pre-war norms (De Toni, 1954), and 2 to 3 years behind the current international references (NCHS Growth Charts, 1976).

At national level, the National Research Council conducted in 1949 a study on about 14,000 0-11 year old children, distributed over 15 regions, the sample being stratified according to geographic areas and socioeconomic background (Bacchetta, 1951). The survey revealed that the child was growing poorly and that the rate of growth was closely associated with latitude, the Northern and Centre children being taller and heavier at all ages and in both sexes than

ECOLOGICAL ENVIRONMENT

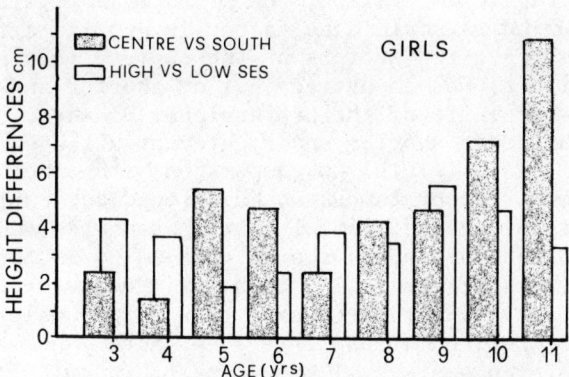

Fig. 2. Geographical and socioeconomic differences in growth of a group of 3-11 year old Italian girls measured in 1949-1950. Children from the central regions and of high socioeconomic status (SES) are taller at all ages than southern children and children of low SES (Based on data from Bacchetta, 1951).

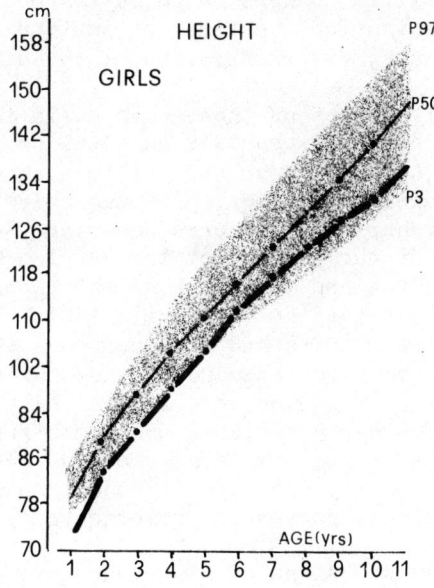

Fig. 3. Mean height of Italian girls measured in 1949-1950 (Based on data from Bacchetta, 1951). The fine line and the stippled area represent the reference standard (P_3, P_{50}, P_{97}; NCHS 1976).

those from the South and the Islands (Fig. 2). Besides and independently of the geographic derivation, the socioeconomic status represented the most important discriminant factor in growth performance. The child from less favoured sectors of the society (unemployed, disabled, welfare recipients) was on average 4.1 cm shorter and 1.6 kg lighter than the well-off child of the professional classes. On the average, the Italian child was shorter and lighter at all ages than the reference (De Toni, 1968). To allow comparison with other sets of data collected in more recent times, we have found desirable to reevaluate these data (Fig.3) against actual growth standards (NCHS Growth Charts, 1976). The comparison showed that 20 to 30% of the children less than 5 years of age fell 3 cm below the reference stature. The average deficit in stature was 5 to 9 cm for males and 4 to 10 cm for females between 4 and 11 years of age. The weight deficit (80% of the reference or less) was present in 23% of the children below 2 years, but only in 12 to 16% up to 9 years, beyond which the deficit was again more common, affecting almost one-third of the sample.

A rather similar picture was still present as late as the '60s, in spite of the beginning of a rapid and intense period of socioeconomic expansion. With the progressive acceleration of the economic growth of the country, the quality of life underwent generalized qualitative and quantitative changes, with better housing conditions, improved social organization, better distribution of basic health services and assistance, diversification of the diet, etc.

During the early stages of these changes in the environment, a national study conducted between 1959 and 1968 on a representative sample of about 100,000 6-11 year old children showed that they still grew less than children from other, presumably well-fed, populations in countries with higher income (Ferro-Luzzi and Sofia, 1967). This study also showed that the child from the rural areas was markedly shorter (Fig.4), lighter and thinner than the urban child and had an inferior nutritional status (Ferro-Luzzi, 1966). The major signs of overt nutritional deficiencies had disappeared with the notable exception of iodine deficiency (endemic goiter, Baschieri et al., 1978). The attenuation, but persistence, of clinical signs of nutritional deficiencies, their higher prevalence among the rural children, the still marginal food availability, confirmed the less than optimal nutritional status of the Italian population and suggested the contribution of the nutritional factor to the observed growth retardation.

The situation showed signs of a profound evolution some years later. Cialfa (1982) calculated on the basis of the national Food Balance Sheets that the Italians were consuming in 1978-1980 almost twice as much sugar, and vegetables and fruits, as compared to the 1954-1956, about twice as much fats and oils, and almost three times as much meat. Milk and dairy product consumption had passed from nearly 62 kg/pc/pa to about 96 kg/pc/pa in 1978-1980. Energy availa-

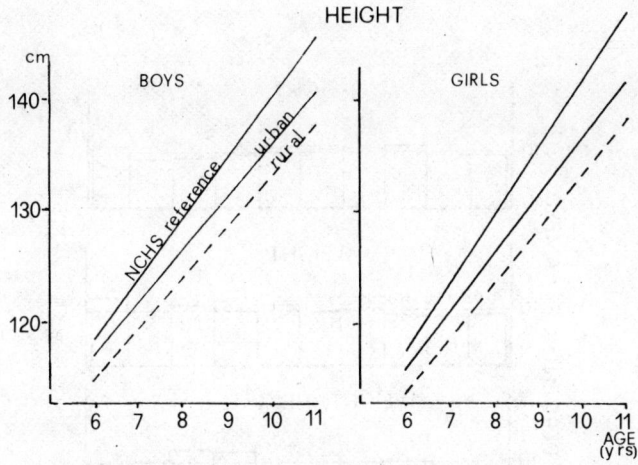

Fig. 4. Regressions of height on age for urban (solid line) and rural (broken line) Italian children from 6 to 11 years (based on data from Ferro-Luzzi and Sofia, 1967).

bility was 3250 kcal/pc/pd, largely exceeding the theoretical national requirement for energy, calculated to be 2450 kcal/pc/pd. Protein availability was 100 g/pc/pd, a 59% increase above the 1951-1960 values. GNP had increased in 1980 to 5,913,000 Liras per year per inhabitant; however, due to the massive inflation, the real increase over the 1951 GNP when depurated of the price component is about three fold (ISTAT, 1981).

As the Rofrano study was repeated in 1979-1980, it can provide a measure of how this revolutionary improvement affected the more deprived areas (Cresta et al., 1982). The study revealed that illiteracy had dropped vertically; less than I child in 4 had intestinal parasites, and none more than 2 different species at the same time. All houses had running water and most of them were endowed with hygienic facilities. The diet met the recommendations for energy and for most of the nutrients and the children were growing better than their counterparts of about 25 years before. Stature had increased for both males and females of about 4% at ages 6 to 15. Weight increased over the 1954 values 7% for boys between 6 and 10 years, and 9% for the girls of the same age. At ages 10-15, the increase in weight over the 1954 values, was 13% for boys and 18% for females. The Rofrano child has not yet reached the international standards for height.

This local evolution was confirmed by wider studies conducted over the Country, showing that the dramatic improvement of environmental conditions had promoted an increase in the body dimensions of

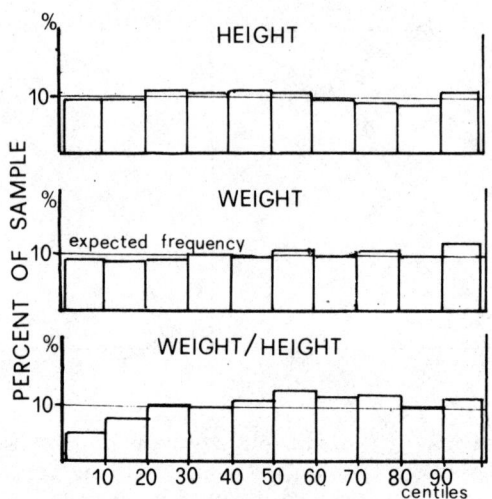

Fig. 5. Centile distribution of growth parameters of 8077 three-to-five yr old Italian children (Redrawn from Ferro-Luzzi and Mariani, 1981).

the population. A cross-sectional survey was carried out in the second half of the '70s by the National Institute of Nutrition on the growth and nutritional status of the preschool child, age 3 to 5, in 8 different areas. Preliminary results have appeared in the literature (Ferro-Luzzi and Mariani, 1981). The study covered about 8000 children in urban and rural areas selected to represent the life styles and socioeconomic realities prevailing in Italy. The results suggest that some of the factors that had previously been associated with growth failure have changed, others have become attenuated, others finally have acquired a new priority. The conventional indicators of physical growth (height, weight for height) proved that the study sample was growing in a satisfactory way (Fig.5), their mean and distribution being identical to the adopted international standards of reference (NCHS Growth Charts, 1976). No difference worth mentioning was also found in the growth of school age children studied in Rome by Mariani et al. (1978) and in Abruzzi, a low to middle-income region of Central Italy, by D'Amicis et al. (1979), thus confirming the general attenuation of the environmental pressure on the genetic potential for growth of the average Italian child.

A more detailed analysis, however, disclosed that, of the factors that had been previously associated with growth retardation, two and namely urbanization and social class stratification, were

ECOLOGICAL ENVIRONMENT

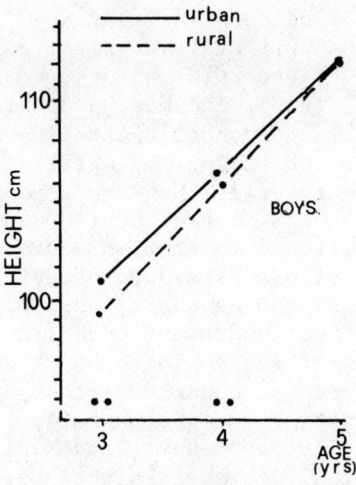

Fig. 6. Mean heights of urban and rural Italian preschool boys.

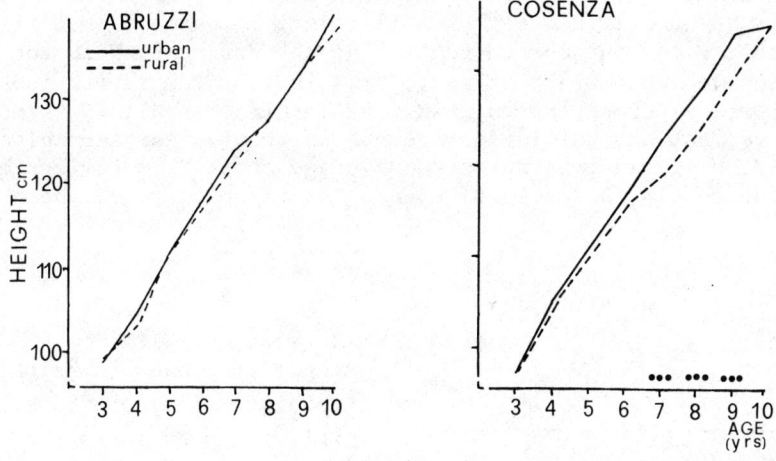

Fig. 7. Heights of children in rural and urban areas. The comparison shows that 3 to 10 year old urban girls from Abruzzi, a region of medium-low income have the same height as their rural peers; in Cosenza, an undervelopped province of the South, urban girls are significantly taller at 7-9 years of age than their rural peers (Redrawn from Ferro-Luzzi and Mariani, 1981; D'Amicis et al., 1979; Maiale et al., 1980).

still operative at a certain level and under certain local conditions.
When the data from the preschool child survey (Ferro-Luzzi and Mariani
1981) were examined from this angle, a small but still significant
difference in size of 3-5 year old children could be seen between
urban and rural areas (Fig.6). The disaggregation at a regional level
of the data on preschool and school age children revealed that the
situation was rather heterogeneous, as marked differences could still
be found in certain areas, while they had practically disappeared
from others. Thus in Abruzzi, a region of medium-low income, the rural
and urban children were found to grow in a similar way (Fig.7), and
both grew to the international standard of reference, from 3 to 10
years of age (Ferro-Luzzi and Mariani, 1981; D'Amicis et al., 1979).
In Cosenza, one of the most underdeveloped Italian Provinces, where
the somatic development of the children had been found to still lag
behind the norm of reference, a marked effect of urbanization could
still be detected, affecting the growth pattern of the rural 7 to 9
year old children (Fig.7) (Ferro-Luzzi and Mariani, 1981; Maiale et
al., 1980). It is difficult to explain this delayed appearance of a
differential growth pattern and further analysis are needed. As the
data are cross-sectional, an explanation would be that the older rural
child could have been exposed, in its earlier infancy, to a different
and less favourable environment than its younger sibling, and has not
yet been able to catch-up when, recently, the conditions have improved.
An alternative explanation might be found in the participation
of the older rural child to physically demanding agricultural duties
that might have proved detrimental to its growth performance (Malina,
1979). We have no evidence that this is still the case, but it was
undoubtedly a common observation in the not far past. When the present
day Cosenza child (Maiale et al., 1980) is compared to its peer studied
about 15 years before (Ferro-Luzzi and Termine, 1963), it is possible
to observe that a sensible improvement in growth has taken place,
the children of today being on the average about 2 cm taller than
those measured in 1963 (Fig.8). Since the improvement has been of

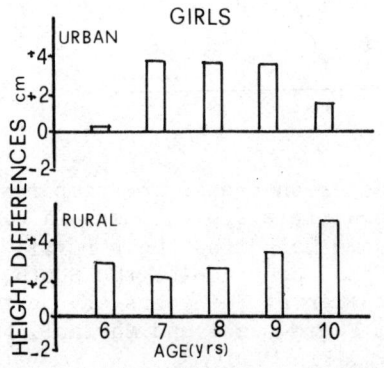

Fig. 8.

Mean differences in height
(absolute values) of Cosenza
girls measured in 1977 over
the height of their peers
measured in 1963, showing
that present day girls, both
urban and rural, are taller
(from Maiale et al., 1980).

similar magnitude under urban and rural conditions, the difference between the two environments has remained practically unmodified.

Urban/rural differences in body dimensions have been described in many other parts of the worls (see before); the quality of life and the societal organization have been invoked as possible factors contributing to the genesis of the phenomenon. Selective migration, the rate of sexual maturation and the level of physical activity have also been considered to cause the differential rates of growth. Finally, some of the differences have been explained on the basis of the higher level of heterosis of the urban population. Earlier studies indicated that at the end of the 18th century the rural areas offered a less stressful environment than cities, in coincidence with the earlier phases of the industrialization process and the concomitant uncontrolled exploitation of the cheap labour of women and young children, the long working hours, overcrowding, etc. (Malina, 1979). At the turn of the century, the cities started to provide better living conditions and, after a long slow transitional period, the situation was fully reversed, the cities presenting a more favourable environment characterized by improved social care and public health facilities, good housing, better sanitation, diversified diets, access to educational facilities, etc. Under these conditions it is not surprising that the urban child fared better than his rural counterpart, enjoyed a better nutritional status and grew taller and heavier (Tanner and Eveleth, 1976). In the recent literature it is even possible to find the further stages of this developmental cycle, with disappearance of the disparity in body size between urban and rural children in countries enjoying a high level of affluency (Van Wieringen et al., 1971). This phenomenon may be attributed to a lower level of endogamy in the rural areas as consequence of the increased mobility of the population, to the slowing down of the increase in body size of the urban child who has in the mean time approached or attained full expression of his growth potential, and to the progressive environmental enrichment of the rural *milieu*.

The urban/rural contraposition suffers from the fact that it does not control for socioeconomic conditions, the elite urban child being lumped together with the slum child, and the affluent farmer with the landless seasonal agricultural labourer. When the present time Italian data (Ferro-Luzzi and Mariani, 1981; D'Amicis et al., 1979; Mariani et al., 1978) were analyzed for socioeconomic differences in growth, using paternal occupational status as proxy indicator, the results disclosed that the difference in stature between the lower and the higher socioeconomic classes was greater and more significant than the urban/rural difference. At the preschool age, the difference between the child of professionals and the working classes was about 2 cm (Fig.9) (Ferro-Luzzi and Mariani, 1981). In areas like the Abruzzi, where no differences in growth existed between urbans and rurals, the difference in stature associated between high and low social classes ranged, at ages 6-10, between 2 and 3 cms (D'Amicis

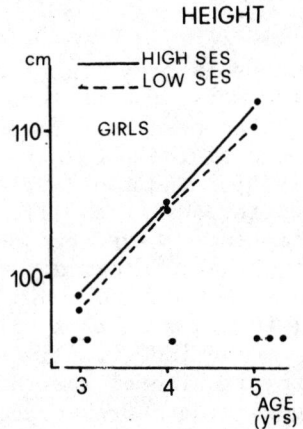

Fig. 9. Socioeconomic differences in height of 3 to 5 year old Italian girls of high and low socioeconomic status (Redrawn from Ferro-Luzzi and Mariani, 1981).

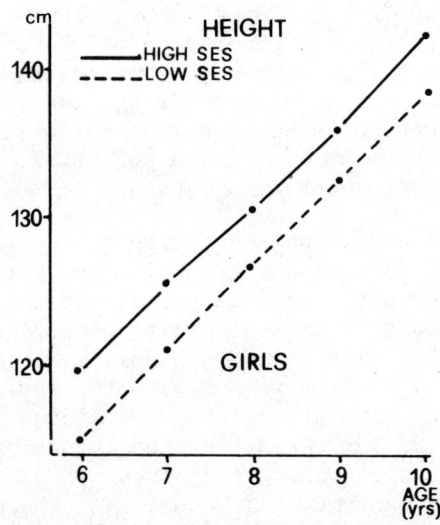

Fig. 10. Socioeconomic differences in height of 6 to 10 year old Roman girls of working (Low SES) and professional (High SES) class (Redrawn from Mariani et al., 1978).

et al., 1979; Ferro-Luzzi and Mariani, 1981). The stature of working class Roman children, studied longitudinally over years, was found to be systematically shorter, about 4 cm, than that of the children of professional classes (Mariani et al., 1978) (Fig. 10). The difference was already present at the age of 6 and persisted unmodified up to age 10, implying similar growth velocities.

It is not easy to explain the meaning and causes of these differences. In the case of the Roman child, the dietary intakes, carefully monitored over the 10 years of duration of the study, did not reveal any class difference in nutrient and energy consumption worth mentioning. The same results were obtained in other groups with similar differences in growth performance. These data appear to suggest that the nutritional factor does no more account for the growth retardation still observed in Italy, and that other factors must be invoked. The identification of these factors and the evaluation of their contribution to the differential growth performance of the Italian child needs further investigation to separate and control for several confounding variables and for starting to unravel and understand the dynamics that regulate growth in Italy today.

REFERENCES

Abbott Seagraves, B., 1977, The Malthusian proposition and nutritional stress: Differing implications for man and for society, in: Malnutrition, behavior and social organization, L.S. Greene, ed., Academic Press, New York.

Acheson, R.M., 1959, Effects of starvation, septicaemia and chronic illness on the growth cartilage plate and metaphysis of the immature rat, J. Anat., 93: 123.

Acheson, R.M. and Fowler, 1964, Sex, socio-economic status, and secular increase in stature: a family study, Brit. J. Prev. Soc. Med., 18: 25.

Alford, C.A., Fort, J.W., Blankenship, W.J., Cassady, G. and Benton, J.W., 1969, Subclinical central nervous system disease of neonates: a prospective study of infants born with increased levels of IgM, J. Pediat., 75: 1167.

Ariza Macias, J., Luna-Jaspe, H., Rueda Williamson, R., Mora Parra, J. and Fardo Telliz, F., 1973, Cross sectional study of growth of Colombian children from different socio-economic classes during their first six years of life, Ecol. Food Nutr., 2: 61.

Aschcroft, M.T., Heneage, P. and Lowell, H.G., 1966, Heights and weights of Jamaican school children of various ethnic groups, Am. J. Phys. Anthrop., 24: 35.

Bacchetta, V., 1951, Alimentazione e stato di nutrizione dei bambini italiani dopo la guerra, La Ricerca Scientifica, suppl., CNR.

Bancroft, J., 1933, The conditions of fetal respiration, Lancet, 225: 1021.

Baschieri, L., Costa, A. and Basile, A., ed., 1978, Il gozzo, L.Pozzi Roma.
Beargie, R.A., James, V.L. and Greene, J.W., 1970, Growth and development of small-for-date newborns, Ped. Clins. N. Amer., 17: 159.
Beisel, W.R., 1975, Nutrient wastage during infection, in: Proc. 9th int. Congr. Nutrition, Mexico, 2: 160 (Karger, Basel).
Blanco, R.A., Acheson, R.M., Canosa, C. and Salomon, J.B., 1974, Height, weight and lines of arrested growth in young Guatemalan children, Am. J. Phys. Anthrop., 40: 39.
Bogin, B. and McVean, R.B., 1981, Body composition and nutritional status of urban Guatemalan children of high and low social economic class, Am. J. Phys. Anthrop., 55: 543.
Butzina, R., Jusic, M., Sapunar, J. and Milanovic, N., 1980, Zinc nutrition and taste acuity in school children with impaired growth, Am. J. Clin. Nutr., 33: 2262.
Campbell, S., 1974, Physical methods of assessing size at birth, in: Size at birth. Ciba Foundation Symposium 27 - Elsevier, Excerpta Med., North Holland.
Chabes, A., Perada, J., Hyams, L., Barrientos, N., Perez, J., Campos, L., Monroe, A. and Mayorga, A., 1968, Comparative morphometry of the human placenta at high altitude and sea level: I The shape of the placenta, Obs. and Gyn., 31: 178.
Chandra, R.K. and Matsumura, T., 1979, Ontogenic development of the immunesystem and effects of fetal growth retardation, J. Perinat. Med., 7: 279.
Chandra, R.K., 1979, Nutritional deficiency and susceptibility to infection, Bull. Wrld. Hlth. Org., 57: 167.
Cialfa, E., 1982, I consumi alimentari, Agricoltura, 109: 15.
Cresta, M. and Fiorentini, M., 1956, Rilievi somatometrici sulla popolazione infantile di una zona depressa, La Clinica Pediatrica, 38: 1.
Cresta, M., Barberini, G., Calandra, P.L., Cialfa, E., De Majo, A.M., Gradoni, L., Gramiccia, M., Mariani, M., Passarello, P., Pozio, E., Pulicani, A.R., Ricci, M. and Vecchi, F., 1982, Rofrano: 1954-1980: Studio biologico su 26 anni di storia di una comunità montana dell'Italia Meridionale (Prov. di Salerno), Quad. Nutr., Monograph. Series no.2, Ntl. Inst. Nutr., ed.
D'Amicis, A., Ferrini, A.M., Ferro-Luzzi, A., Maiale, G. and Perozzi, G., 1979, Accrescimento e stato di nutrizione di bambini abruzzesi di 6-10 anni nelle diverse situazioni socio ambientali, 12th Gen. Meet. SINU, L'Aquila.
De Toni, G., 1968, Auxologia: Auxologia postnatale fisiologica, p.31 Minerva Medica, Roma.
Eastman, N.J., 1930, Foetal blood studies: I. The oxygen relationships of umbilical cord blood at birth, Bull. John Hopkins Hospital, 47: 221.
Ferro-Luzzi, A., 1962, Lo stato di nutrizione delle comunità scolari italiane. I^ Rassegna critico-sintetica sul livello nutrizionale della popolazione italiana, Quad. Nutr., 22: 29.

Ferro-Luzzi, A., 1967, Reddito e stato di nutrizione in Italia, Quad. Nutr., 27: 124

Ferro-Luzzi, A., D'Amicis, A., Ferrini, A.M. and Maiale, G., 1979, Nutrition, environment and physical performance of preschool children in Italy, Biblthca Nutr. Dieta, 27: 85.

Ferro-Luzzi, A. and Mariani, A., 1981, Nutritional deficiencies in preschool and prepuberal children, Biblthca Nutr. Dieta, 30: 30.

Ferro-Luzzi, A., 1982, Meaning and constraints of energy intake studies in free-living populations, in: Energy and effort. Taylor & Francis, London.

Ferro-Luzzi, G. and Sofia, F., 1962, Fabbisogno calorico e livello nutrizionale della popolazione italiana, Notiz. Amm. Sanit.,122.

Ferro-Luzzi, G. and Termine, L., 1963, Lo stato di nutrizione delle comunità scolari italiane. IX: Lo stato di nutrizione degli scolari della provincia di Cosenza, Quad. Nutr., 23: 214.

Ferro-Luzzi, G., 1966, The nutritional status of the rural population in 16 Italian communities. Quad. Nutr., 26: 94.

Ferro-Luzzi, G. and Sofia, F., 1967, Rapporto su altezze, pesi ed altri indici di stato di nutrizione dei bambini italiani (età 6-11 anni), Quad. Nutr., 27: 269.

Frisancho, A.R., Garn, S.M. and Ascoli, W., 1970, Childhood retardation resulting in reduction of adult body size due to lesser adolescent skeletal delay, Am. J. Phys. Anthrop., 33: 325.

Frisancho, A.R., Sanchez, J., Pallardel, D. and Yanez, L., 1973, Adaptive significance of small body size under poor socio-economic conditions in Southern Peru, Am. J. Phys. Anthrop., 39: 255.

Frisancho, A.R., Klayman, J.E. and Matos, J., 1974, Influence of maternal nutritional status on prenatal growth in a Peruvian urban population, Am. J. Phys. Anthrop., 46: 265.

Frisancho, A.R., Guire, K., Babler, W. Borkan, G. and May, A., 1980, Nutritional influence on childhood development and genetic control of adolescent growth of Quechuas and Mestizos from the Peruvian Lowlands, Am. J. Phys. Anthrop., 52: 367.

Frisancho, A.R., 1980, Role of calorie and protein reserves on human growth during childhood and adolescence in a mestizo Peruvian population, in: Social and biological predictors of nutritional status, physical growth and neurological development, L.S.Greene and F.E. Johnston, ed., Academic Press, New York.

Frontali, G., 1950, Razione alimentare e stato di nutrizione del bambino italiano dopo la guerra, Quad. Nutr., 11: 133.

Garn, S.M. and Rohmann, C., 1966, Interaction of nutrition and genetics in the timing of growth and development, Pediatric. Clin. North America, 13: 353.

Garn, S.M., Bailey, S.M. and Cole, P.E., 1977, Similarities between parents and their adopted children, Am. J. Phys. Anthrop., 45: 539.

Garn, S.M., Shaw, H.A. and McCabe, K.D., 1978, Effect of socioeconomic status on early growth as measured by three different indicators, Ecol. Food. Nutr., 7: 51.

Goldstein, H., 1971, Factors influencing the height of seven year old children. Results from the National Child Development Study, Human Biol., 43: 92.

Gopolan, C., Swamintham, M.C., Kumari, V.K.K., Rao, D.H. and Vijayaraghavan, K., 1973, Effect of calorie supplementation on growth of undernourished children, Am. J. Clin. Nutr., 26: 563.

Graham, G.G., Creed, H.M., McLean, W.C., Kallman, C.H., Rabold, J. Mellits, E.D., 1981, Determinants of growth among poor children: nutrient intake-achieved growth relationship. Am. J. Clin. Nutr., 34: 539.

Greiner, T. and Latham, M.C., 1981: Factors associated with nutritional status amoung young children in St. Vincent., Ecol. Food Nutr., 10: 135.

Haas, J.D., 1974, Variation in the pattern of sexual dimorphism of infants under environmental stress., Am. J. Phys. Anthropol., 41: 483.

Haas, J.D., Baker, P.T. and Hunt Jr., E.E., 1977, The effects of high altitude on body size and composition of the newborn infant in southern Peru. Hum. Biol., 49: 611.

Haas, J.D., 1980, Maternal adaptation and fetal growth at high altitude, in: Social and biological predictors of nutritional status, physical growth and neurological development, L.S. Greene and F.S. Johnston, ed., Academic Press, New York.

Haas, J.D., Frongillo, E.A., Stepick, C.D., Beard, J.L. and Hurtado, L.G., 1980, Altitude, ethnic and sex difference in birth weight and length in Bolivia, Hum. Biol., 52: 459.

Habicht, J.P., Lechtig, A., Yarbrough, C. and Klein, R.E., 1974, Maternal nutrition, birth weight and infant mortality, in: Size at birth. CIBA Symposium 27 (new series) Elsevier Excerpta Med., North Holland.

Habicht, J.P., Martorell, R., Yarbrough, C., Malina, R.M. and Klein, R.E., 1974, Height and weight standards for preschool children. How relevant are ethnic differences in growth potential?, Lancet, i: 611.

Halsted, J.A., Ronaghy, H.A., Abadi, P., Haghshenass, M., Amirhakemi, G.H., Barakat, R.N. and Reinhold, J.G., 1972, Zinc deficiency in man. The Shivaz experiment, Am. J. Med., 53: 277.

Hambidge, K.M., Hambidge, C., Jacobs, M. and Baum, J.D., 1972, Low levels of Zinc in hair, anorexia, poor growth and hypogeusia in children, Pediat. Res., 6: 868.

Hambidge, K.M., Walravens, P.A., Brown, R.M., Webster, J., White, S., Anthony, M. and Roth, M.L., 1976, Zinc nutrition of preschool children in the Denver Head Start program, Am. J. Clin. Nutr., 29: 734.

Harris, A.A., 1931, Lines of arrested growth in long bones in childhood: correlation of histological and radiographic appearances in clinical and experimental conditions, Brit. J. Radiol., 4:562.

Hewitt, D., Westropp, C.K. and Acheson, R.M., 1955, Oxford child health survey. Effect of childish ailments on skeletal development, Brit. J. Prev. Soc. Med., 9: 179.

Hewitt, D., 1957, Some familial correlations in height, weight and skeletal maturity, Ann. Human Genet., 22: 26.
Howard, R.C., Lichty, J.A. and Bruns, P.D., 1957, Studies of babies born at high altitude. II Measurement of birth weight, body length and head size, Am. J. Dis. Child., 93: 670.
ISTAT, 1981, Conti economici nazionali anni: 1965-1980. Collana di informazione, Anno 5, n°1.
ISTAT, 1981, Annuario di contabilità nazionale, 9: 1 ed. 1979, Roma.
Johnston, F.E., Borden, M. and MacVean, R.B., 1973, Height, weight and their growth velocities in Guatemalan private school children of high socioeconomic class, Hum. Biol., 45: 627.
Johnston, F.E., Wainer, H., Thissen, D. and McVean, R., 1977, Hereditary and environmental determinants of growth in height in a longitudinal sample of children and youth of Guatemalan and European ancestry, Am. J. Phys. Anthrop., 44: 469.
Johnston, F.E., Newman, B., Cravioto, J., Delicardie, E. and Scholl, T., 1980, A factor analysis of correlates of nutritional status in Mexican children, birth to 3 years, in: Social and Biological predictors of nutritional status, physical growth and neourological development,L.S.Greene & F.E.Johnston,eds.,Acad.Press,New York.
Johnston, F.E.,1981, Physical growth and development and nutritional status: epidemiological considerations. Federation Proc.,40:2583.
Katz, M., Keusch, G.T. and Mata, L., eds., 1975, Malnutrition and infection during pregnancy; determinants of growth and development of the child, Am. J. Dis. Child., 129: 419.
Keller, W., Donoso, G. and De Mayer, E.M., 1976, Anthropometry in nutritional surveillance: a review based on results of the WHO collaborative study on nutritional anthropometry, Nutr. Abs. & Rev., 46: 591.
Lechtig, A. and Mata, L.J., 1971, Cord IgM levels in Latin American neonates, J. Pediatr., 78: 909.
Lechtig, A.,Yarbrough, C., Delgado, H., Habicht, J., Martorell, R. and Klein, R.E., 1975, Influence of maternal nutrition on birth weight, Am. J. Clin. Nutr., 28: 1223.
Lechtig, A., Martorell, R., Delgado, H., Yarbrough, C. and Klein, R.E., 1978, Food supplementation during pregnancy, maternal anthropometry and birth weight in a Guatemalan rural population, J. Trop. Pediatr., 12: 217.
Lederman, S.A. and Rosso, P., 1980, Effects of food restriction on fetal and placental growth and maternal body composition. Growth, 44: 77.
Levinson, F.J., 1977, Morinda: An economic analysis of malnutrition among young children in rural India. Cornell/MIT Intern. Nutrition Policy Series. Cambridge, Massachussets, U.S.A.
Lichty, J.A., Ting, R.Y., Bruns, P.D. and Dyar, E., 1957, Studies of babies born at high altitudes. I Relation of altitude to birth weight, Am. J. Dis. Child., 93: 666.
Lindgren, G., 1976, Height, weight and menarche in Swedish urban school children in relation to socio-economic and regional factors, Ann. Hum. Biol., 3: 501.

Livson, N., McNeill, D. and Thomas, K., 1962, Pooled estimates of parent-child correlations in stature from birth to maturity, Science, 138: 818.

Logie, D.E., McGregor, I.A., Rowe, D.S. and Billewicz, W.Z., 1973, Plasma immunoglobulin concentrations in mothers and new born children with special reference to placental malaria. Studies in Gambia, Nigeria and Switzerland, Bull. Wrld. Hlth. Org., 49: 547.

Lownstein, F.W. and O'Connell, D.E., 1974, Selected body measurements in boys ages 6-11 years from six villages in Southern Tunisia: An International comparison, Hum. Biol., 46: 471.

Maiale, G., D'Amicis, A., Ferrini, A.M. and Ferro-Luzzi, A., 1980, La malnutrizione in Italia: la situazione del bambino cosentino. Proc. 13th Gen. Meet. S.I.N.U., Caserta.

Malcolm, L.A., 1970, Growth and development in New Guinea: a study of the Bundi people of the Madang District, Inst. Hum. Biol. Madang, Monogr. Series, no.1.

Malina, R.M., 1979, Secular changes in size and maturity: causes and effects, in: Secular trends in human growth, maturation and development, A.F. Roche, ed., Monograph of the Society for Research in Child Development n°179, University of Chicago Press.

Mancini, F. and Fiorentini, M., 1957, Indagini precedenti ed orientamenti attuali sui rapporti tra alimentazione e pellagra, Notiz. Ammin. Sanit., 10: 1.

Marchione, T.J. and Prior, F.W., 1980, The dynamics of malnutrition in Jamaica, in: Social and Biological predictors of nutritional status, physical growth and neurological development, L.S. Greene and F.E. Johnston, ed., Academic Press, New York.

Mariani, A., Lancia, B., Migliaccio, P.A. and Sorrentino, D., 1978, Growth, nutritional status and food consumption of Roman school children, in: Auxology: Human growth in health disorder, L.Gedda and P. Parisi, ed., 13: Academic Press, London.

Martorell, R., Lechtig, A., Yarbrough, C., Delgado, H. and Klein, R.E., 1978, Energy intake and growth in an energy deficient population, Ecol. Food and Nutr., 7: 147.

Martorell, L., Yarbrough, C., Klein, R.E. and Lechtig, A., 1979, Malnutrition, body size and skeletal maturation: interrelationships and implications for catch-up growth, Hum. Biol., 51: 371.

Martorell, R., 1980, Interrelationships between diet, infectious disease and nutritional status, in: Social and biological predictors of nutritional status, physical growth and neurological development, L.S. Greene and F.E. Johnston, eds., Academic Press, New York.

Martorell, L., Yarbrough, C., Lechtig, A., Habicht, J.P. and Klein, R.E., 1975, Diarrheal diseases and growth retardation in preschool Guatemalan children, Am. J. Phys. Anthrop., 43: 341.

Mata, L.J., Urrutia, J.J., Albertazzi, C., Pellecer, O. and Arellano, E., 1972, Influence of recurrent infections on nutrition and growth of children in Guatemala, Am. J. Clin. Nutr., 25: 1267.

Mata, L.J., Kromal, R.A., Urrutia, J.J. and Garcia, B., 1977, Effect of infection on food intake and the nutritional state: perspectives as viewed from the village, Am. J. Clin. Nutr., 30: 1215.
Mazess, R.B., 1975, Human adaptation to high altitude, in: Physiological anthropology, A. Damon, ed., Oxford University Press, New York.
McCance, R.A. and Widdowson, E.M., 1974, The determinants of growth and form, Proc. R. Soc. Lond., B 185: 1.
McClung, J., 1969, Effects of high altitude on human birth. Harvard University Press, Cambridge.
McDonald, E., Pollitt, E., Mueller, W., Hsueh, A.M. and Sherwin, R., 1981, The Bacon Chow study: maternal nutritional supplementation and birth weight of offspring, Am. J. Clin. Nutr., 34: 2133.
Meyer, M.B., Jonas, B.S. and Tonascia, J.A., 1976, Perinatal events associated with maternal smoking during pregnancy, Am. J. Epidem., 103: 464.
Meredith, H.V., 1975, Relation between tobacco smoking of pregnant women and body size of their progeny, a compilation and synthesis of published studies, Hum. Biol., 47: 451.
Metcoff, J., 1981, Maternal nutrition and fetal outcome, Am. J. Clin. Nutr., 34: 708.
Monod Cassidy, C., 1980, Benign neglect and toddler malnutrition, in: Social and biological predictors of nutritional status, physical growth and neurological development, L.S. Greene and F.S. Johnston, eds., Academic Press, New York.
Morà, J.A., De Parades, B., Wanger, M., De Navarro, L. and Suescum, J., 1979, Nutritional supplementation and the outcome of pregnancy. I: Birth weight, Am. J. Clin. Nutr., 32: 455.
Mueller, W.H., 1976, Parent child correlations for stature and weight among school aged children: a review of 24 studies, Hum. Biol., 48: 379.
Naismith, D.J., 1969, The foetus as a parassite, Proc. Nutr. Soc., 28: 25.
NCHS Growth Charts, 1976, United States Department of Health, Education and Welfare, Public Health Service, Health Resources Administration, Rockville, Md. HRA 76- 1120, 25: 3
Ounsted, M., 1971, Foetal growth, in: Recent adv. in Paediatrics, D. Gairdner and D. Hull, eds., Churchill, London.
Owen, G.M., Kram, K.M., Garry, P.J., Lowe, J.E. and Lubin, A.H., 1974, A study of nutritional status of preschool children in the United States, 1968-1970, Pediatr., 53: 597.
Prasad, A.S., Micale, A., Farid, Z., Sandstead, H. and Schulert, A.R., 1963, Zinc metabolism in patients with the syndrome of iron deficiency anemia, hepatosplenomegaly, dwarfism and hypogonadism, J. Lab. Clin. Med. 61: 537.
Rosso, P., 1977a, Maternal-fetal exchange during malnutrition in the rat. Transfer of glucose and methyl (α-D-U-14C glucose) pyranoside, J. Nutr., 107: 2006.
Rosso, P., 1977b, Maternal-fetal exchange during malnutrition in the rat. Transfer of α-amino isobutyric acid, J. Nutr., 107: 2002.

Rosso, P., 1981, Nutrition and maternal fetal exchange, Am. J. Clin. Nutr., 34: 744.
Rush, D., Stein, Z. and Susser, M., 1980, A randomized controlled trial of prenatal nutritional supplementation in New York City, Pediatr., 65: 683.
Russell, M., 1976, Parent-child and sibling-sibling correlations of height and weight in a rural Guatemalan population of preschool children, Hum. Biol., 48: 501.
Schutte, J.E., 1980, Growth differences between lower and middle income Black male adolescents, Hum. Biol., 52: 193.
Soysa, P.E. and Jayasuriya, D.S., 1975, Birth weight in Ceylonese, Hum. Biol., 47: 1.
Stini, W.A., 1972, Reduced sexual dimorphism in upper arm muscle circumference associated with protein-deficient diet in a South American population, Am. J. Phys. Anthrop., 36: 341.
Stini, W.A., 1981, Association of early growth patterns with the process of aging, Federation Proc., 40: 2588.
Tanner, J.M. and Israelsohn, W.J., 1963, Parent-child correlations for body measurements of children between the ages one month and seven years, Ann. Human Genet., 26: 245.
Tanner, J.M., 1963, The regulation of human growth, Child. Dev., 34: 817.
Tanner, J.M. and Eveleth, P.B., 1976, Urbanization and growth, in: Man in urban environments, G.A. Harrison and J.B. Gibson, eds., Clarendon Press, Oxford.
Thomson, A.M., Billewicz, W.Z. and Hytten, F.E., 1968, The assessment of foetal growth, J. Obstet. Gynecol., 75: 903.
Underwood, E.J., 1971, Trace elements in humans and animal nutrition, Academic Press, New York.
Valverde, V., Pivaral, V.M., Delgado, H., Belizan, J., Klein, R.E. and Martorell, R., 1981, Income and growth retardation in poor families with similar living conditions in rural Guatemala. Ecol. Food Nutr., 10: 241.
Van Wieringen, J.C., Wafelbakker, F., Verbrugge, H.B. and de Haas, J.H., 1971, Growth diagrams 1965, Netherlands: second national survey on 0-24 year olds, Groningen, Wolters -Noordhoof.
Walravens, P.A. and Hambidge, K.M., 1976, Growth of infants fed a Zinc supplemented formula, Am. J. Clin. Nutr., 29: 1114.
Waterlow, J.C. and Rutishauser, I.H.E., 1974, Malnutrition in man, in: Early malnutrition and mental development, J. Cravioto, L. Hambraeus and B. Valhlquist, eds., Swedish Nutrition Foundation Symposia 12, Almquist & Wiksell, Stockholm.
Welon, Z. and Bielicki, T., 1971, Further investigations of parent-child similarity in stature, as assessed from longitudinal data, Hum. Biol., 43: 517.
Wolanski, N., 1970, Genetic and ecological factors in human growth, Hum. Biol., 42: 349.

ECOLOGICAL ENVIRONMENT

ENVIRONMENT AND PHYSICAL GROWTH:
RANDOM COMMENTS ON FERRO-LUZZI'S OVERVIEW

R. J. Rona

Department of Community Medicine
St. Thomas's Hospital Medical School
London, SE1 7EH, England

The initial dilemma in discussing such a broad topic as the impact of the environment on physical growth is to define the boundaries of the review. A unifying theory that explains all the observed phenomena and at the same time gives guidance for intervention in communities has not been reached. Those of us working in departments of public health use physical growth as an outcome measure to assess the effects of modifications in the environment on the nutritional status of children. Researchers working in the field have misgivings about this undertaking. Our experience in descriptive studies is immense. Sound and generally consistent knowledge of patterns of growth and the influence of the environment have been obtained from cross-sectional studies. However, our understanding of the relation between growth and the environment over time is less satisfactory. Similarly we are less clear about the effect of nutritional efforts on growth and other intervention programmes aimed at improving the nutritional status of the community.

Ferro-Luzzi's paper gives a comprehensive overview of many of the factors found to be associated with physical growth. I will only discuss a few of the topics she covered. These are mainly concerned with the social environment. I want to make the point that most of my experience in this field has been obtained in the U.K. and this may colour my comments. It is possible that many of the factual examples may be less relevant to developing countries, but some of the principles involved could be helpful. I will review aspects of the social environment and growth that are not well understood and may provoke some discussion. It is not my intention to give an exhaustive account but to suggest some avenues that have been less well explored.

Social class and growth
─────────────────────

In the U.K. and many other countries the association between social class and height has been shown over and over again in studies that vary in design and sample size[1-4]. The tendency for children in lower social classes, especially the unskilled social class, to be shorter than those in non-manual social classes has been shown in most surveys in the world with the exception of Sweden[5]. Our knowledge of how these differences arise over time is more modest. Ferro-Luzzi commented that these differences are already apparent at birth and increase progressively up to the age of seven years indicating an accumulative environmental effect (from Garn et al., 1978)[6]. This would suggest that any intervention aimed at children under seven years may halt this trend. It is not easy to find appropriate longitudinal analyses, but Smith et al.[7] analysed annual increments of height in children aged 5 to 10 years old and concluded that there was little change in the relative heights of children from different social classes between these ages. Thus most of the differences between social classes appeared to be established before children entered school, that is before the age of five. Results from a preschool survey in the U.K. suggest that most of the differences were already established at age two[8]. These different results in the U.S.A. and the U.K. can arise from the study design, the type of treatment of the data or from real differences between countries. A longitudinal analysis covering a longer period than a year is to be welcomed. If methodological weaknesses were left aside and the between countries differences were real this apparent contradiction could be an stimulus to furthering our knowledge of how the environment and growth are related. This is clearly to be distinguished from surveillance of nutritional status where undoubtedly unequivocal results would be more helpful for suggesting advice, and these results should be carefully selected for monitoring of groups of children. The opportunity for intervention would be greatly enhanced if we knew the time when the effects of environment on growth are the greatest and when they are minimal during childhood and adolescence.

We also need a better understanding of the meaning and scope of the variables used to assess the influence of the environment on growth. Most of independent variables are highly complex in themselves and the interpretation of results are manifold, contradictory and sometimes unhelpful. It is to be feared that with the impressive advances in computer technology it will be always easier to include more independent variables in the model rather than to attempt to understand the meaning behind the associations or lack of them between these variables and growth. The inclusion of father's social class is an appropriate example to illustrate how knowledge of the influence of the environment on growth could be increased. Social class is a proxy measure not only of family income and the ability to acquire goods but also of life style. Behavioural patterns vary according to social class. Even this distinction of two components in the social class variable is very rough indeed. Life style in particular is made up by a complex set of behaviours. In spite of this an initial approach would be to attempt to distinguish the effects of these two major components of social class, i.e., income and life style on growth. If these were mutually exclusive intervention strategies could be better designed and evaluated. A first analysis would be to assess growth according to social class at different levels of income. The aim of this analysis would be to study the life style component since it can be assumed that after excluding the income component what is left is life style. Table 1 gives the standardised height by social class for two income levels in England and Scotland in 1972. In both countries income and social class were independently associated with height although in non-manual social classes there is almost no difference in height in the two income levels. Unfortunately this type of analysis may have some drawbacks as income and social class are correlated and numbers in some subgroups are very small. Also since parents' income information in this study was not checked with official records some high incomes in low social class and low incomes in high social classes are a little suspect. If the social class trend was not due to income one would be tempted to hypothesise that life style was the major component of the association between social class and height. Health education and welfare support would, in this case, be a possible line of action. On the other hand if all the height differences between social classes were due to income we would be more interested in welfare support. Even if both components of social class are found to be significantly associated with height, as we found in England and Scotland, the information is useful to prevent us from very high expectations from a simple intervention measure. Such evidence does not, however, preclude the need to evaluate any implemented intervention strategy.

The influence of life style on nutritional status has been poorly researched. Thus recent reports in the Symposium of the American Institute of Nutrition[9] of studies of cultural differences and their association with growth are particularly welcome. It is uncertain whether this type of analysis will come up with very striking findings. There are, however, data indicating that the association between parental smoking and growth of children is not entirely due to maternal smoking during pregnancy[10]. In our study the number of smokers at home was significantly associated with children's height after making adjustments for birth weight, parents' height, family size and father's social class. This suggests that family life style not directly related to food intake or prenatal influences may affect the growth of children.

TABLE 1: Age and sex standardised height by father's social class and income

Social Class	No.	Income Low Mean	SD	No.	High Mean	SD
		ENGLAND				
Non-manual classes	76	0.23	0.95	1396	0.21	0.96
Skilled manual workers	187	-0.18	1.02	2067	-0.02	1.01
Semiskilled manual workers	134	-0.20	1.09	633	-0.02	0.93
Unskilled manual workers	152	-0.60	1.11	183	-0.19	0.99
		SCOTLAND				
Non-manual classes	23	0.04	0.99	287	0.18	0.95
Skilled manual workers	122	-0.50	1.04	709	0.08	0.97
Semiskilled manual workers	46	-0.17	0.81	179	-0.08	0.99
Unskilled manual workers	45	-0.61	0.80	75	-0.05	0.99

That passive smoking and growth are causally related rests on rather uncertain evidence but it shows an interesting aspect of the association between the environment and growth and contrasts with the failure to show an association between food intake and growth in developed countries. Because smoking behaviour is more prevalent in lower social classes in the U.K. the above finding can be included in a model to explain different aspects of life style associated with different social classes.

In summary there are two areas where research on social factors and growth are scarce. Firstly longitudinal studies of children to determine more precisely the effect of social factors on growth over time. Secondly better understanding is needed of the effect of individual components of complex variables such as social class on height.

Assessing the impact of economic recession and natural disasters on children's growth

Governments should be interested in assessing the possible effects of economic crisis or natural disasters on growth. This is only possible if there is surveillance of growth. Two extreme reactions to crises can be distinguished. In the Ogaden[11], for example, a drought may affect almost instantaneously the nutritional status of children living there. In Western European countries, on the other hand, the ability to adapt to, for example,

an economic crisis is very high indeed. In the Ogaden there is
little room for manoeuvre, in the latter there is ample. A lot
more time would be needed in Western European Countries before the
effects of sharp changes in economic circumstances would alter the
nutritional status of their communities. It is worth noting that
even in the economic depression of the thirties there were few
indications that the nutritional status of children was affected.
Palmer's study in the 1930s[13] failed to show an association between weight and social conditions in a comparison of children's
nutritional status before and during the depression in the U.S.A.
Baker and coleagues[14] concluded that during the depression in the
thirties there was a steady increase in the height of eight and
nine year-old children in the Rhondda Valley (Wales) despite the
severe social conditions prevailing in the mining industry, the
main economic activity at that time. Some deficiencies in the
above studies can be noted such as the use of weight and not
height in Palmer's study and the lack of comparison by degrees of
poverty in Baker et al.'s analysis. But still the adaptation to
adverse condition in these societies seem quite large with the
exception of war situation illustrated by Dahlman and Petersen[15].
In the U.K. an average of 18% of a person's income is devoted to
food and there is minimal variation in the per capita expenditure
on food by income of the principal earner in the household[16].
Furthermore several studies have failed to find large differences
in food intake in preschool children in the U.K.[17,18]. Information has also been obtained that showed children from lower social
classes to consume more energy and protein than those from higher
social classes[18]. These findings, however, do not indicate that
as a result of economic crisis, nutritional status will not
eventually deteriorate. But they do suggest that there may be a
time lag before these changes occur. An example of the association
between differences in children's height by father's employment
status (whether gainfully employed) according to cohorts defined
by year of birth may illustrate the point. In the U.K. unemployment rose from 0.8M in 1972 to 1.8M registered unemployed in
1980. As unemployment indicates a serious deterioration in a
family's standard of living a consequent deterioration of children's
height was expected. The cross-sectional analysis based on data
collected in 1972 showed an association between height and father's
employment status[4]. In Table 2 the results by birth cohorts are
shown for children at mean age six in the National Study of Health
and Growth. In the four early cohorts of the study a marked difference in height was observed in English boys and girls according
father's employment status. This tended to disappear as the
percentage of children whose fathers were unemployed increased.
Thus the association between employment status and height is
marked only in the hard core of children with an unemployed
father, that is the group who were unemployed when the economic
conditions were better. The new contingent that have joined the

TABLE 2: Differences of mean height by father's employment status and percentage of unemployment in birth cohorts at mean age six (England)

Survey	Birth Cohort	BOYS		GIRLS	
		% of Unempl.	Δ Empl-Unempl† (cm)	% of Unempl	Δ Empl-Unempl† (cm)
1972	1966	3.2	2.53	5.0	4.86
1973	1967	2.2	1.39	3.7	3.56
1974	1968	2.3	3.15	2.6	4.04
1975	1969	4.8	3.61	3.1	2.38
1976	1970	5.9	0.66	3.8	4.38
1977	1971	5.6	-0.29	7.1	0.75
1978	1972	5.0	2.60	7.5	1.69
1979	1973	7.2	0.11	4.0	-0.23
1980	1974	6.1	-0.12	6.9	1.15

† Δ Empl-Unempl = Mean height difference between children whose fathers are employed and those who are unemployed.

unemployed sector recently has caused an increase in the average height of the unemployed group. The real test as to whether father's employment status affects children's height will be over a longer period of time. Thus provided that unemployment continues at the 1980 level, any further increase in the gap between the height of children from families whose father is unemployed and those whose father is in full-time employment will indicate that unemployment has affected growth.

In summary swift changes of growth pattern following changes in socio-economic circumstances in Western European countries are not to be expected. It would be helpful to determine how sensitive to environmental factors, economic or natural disasters, is height in countries of different socio-economic level. In Ogaden the pattern could be of instantaneous changes, in Jamaica[12] the changes could be slower while in Western European Countries the pace could be so slow that it will take long before anything is detected.

Degree of urbanisation and growth: an odd finding

Ferro-Luzzi showed that children in rural areas are smaller than those living in cities in Italy but that improvements in socio-economic circumstances since 1950 have eroded this differences. This has been observed in many other countries. The large migration from the countryside to the city has been included as an important factor in the process that contributed to the secular trend of growth. The majority view has been that urbanisation brings increased economic benefits which result in improved nutrition and child care[19]. Malina[20], while subscribing to this view, noted that in Europe at the end of the 19th century and beginning of the 20th century children in rural areas tended to be taller than those from urban areas and no consistent trend is apparent from surveys carried out in the U.S.A. between 1910 and 1972.

In the U.K. the secular trend of growth has been well documented[21]. There is, however, no published evidence showing rural children to be shorter than urban children from any English or Scottish study. On the contrary an inverse association between height and population density in the U.K. has been consistently observed over the last 300 years. Firstly reported by Francis Bacon[22] in the 17th century, it has been confirmed by Galton (1876)[23], Tuxford and Glegg (1911)[24], Paton and Findlay (1926)[25] Douglas and Blomfield (1958)[26] and Foster, Chinn and Rona (1982)[27].

Douglas and Blomfield attributed the rural/city gradient in height to the higher proportion of professionals living in the country. Foster and colleagues, however, showed that the inverse association between degree of urbanisation and height is not eliminated by adjusting for father's social class, number of children in the family, parents' height, child's birth weight and home overcrowding. Thus, at least in the U.K. urbanisation per se cannot be the cause of the observed secular trend of growth. Urbanisation may have been a confounding variable in the process and may reflect the association of this with general improvements in the standard of living during the century in many other countries. The U.K. results of the association between growth and urbanisation are of particular importance because it forces us to be more critical of the more general views about the relationship between growth and environment.

Conclusions

My comments on Ferro-Luzzi's contribution were aimed at discussing the flaws and omissions in research on environment and growth. The community as opposed to the individual has been the centre of this discussion. Auxologists have proclaimed their interest in public health and that physical growth occupies a prominent place in the assessment of nutritional status and wellbeing of communities. More efforts, however, should be devoted to understanding how the environment and physical growth interact over time. This would give information about the opportunities for intervention and help to predict the outcome of such actions. This does not negate our responsibility for evaluating any actions that are introduced.

Even for those aspects of the environment that have been more thoroughly researched there are notorious gaps. If more efforts are not made to fill them the future of auxology would be only the repetition of descriptive studies and surveillance of nutritional status but it will scarcely contribute to further advances in the social and health care of children in the community.

REFERENCES

1. J. W. B. Douglas, and H. R. Simpson, Height in relation to puberty, family size and social class: A longitudinal study. Milbank Mem. Found. Q. 42: 20 (1964).
2. H. Goldstein, Factors influencing the height of seven-years-old children: results from the National Child Development Study, Hum. Biol. 43: 92 (1971).
3. F. J. W. Miller, S. D. M. Court, E. G. Knox, and S. Brandon, "The school years in Newcastle upon Tyne 1952-62 being a further contribution to the study of a thousand families," pp 55-56. Oxford University Press, London, (1974).
4. R. J. Rona, A. V. Swan, and D. G. Altman, Social factors and height of primary schoolchildren in England and Scotland, J. Epidemiol. Community Health, 32: 147 (1978).
5. G. Lindgren, Height, weight and menarche of Swedish urban schoolchildren in relation to socio-economic regional factors, Ann. Hum. Biol. 3: 501 (1976).
6. S. M. Garn, H. A. Shaw, and K. D. McCabe, Effect of socio-economic status on early growth as measured by three different indicators, Ecol. Food Nutr. 7: 51 (1978).
7. A. M. Smith, S. Chinn, and R. J. Rona, Social factors and height gain of primary schoolchildren in England and Scotland, Ann. Hum. Biol. 7: 115 (1980).
8. P. T. Fox, M. D. Elston, and J. C. Waterlow, Pre-school child survey pp 64-84. In "Department of Health and Social Security. Reports on health and social subjects No. 21 Subcommittee on nutritional surveillance". Second Report, H. M. S. O., London (1981).
9. R. S. Weinstein, and C. M. Cassidy, Nutritional anthropology: theory and method. Overview of nutritional anthropology, Federation Proc. 40: 2570 (1981).
10. R. J. Rona, C. du V. Florey, G. C. Clarke, and S. Chinn, Parental smoking at home and height of children, Br. Med. J. 283: 1363 (1981).

11. M. Gebre-Medhin, Nutrition surveillance in developing countries, with especial reference to Ethiopia pp 431-441., In "Human Nutrition Vol. 2 Nutrition and Growth", D. B. Jelliffe and E. F. P. Felliffe, eds. Plenum Press, London (1979).
12. T. J. Marchione, and F. W. Prior, The dynamics of malnutrition in Jamaica pp 201-222 In "Social and Biological Predictors of Nutritional Status, Physical Growth and Neurological Development". L. S. Greene and F. E. Johnston, eds. Academic Press, London (1980)
13. C. E. Palmer, Growth and the economic depression, Public Health Reports, 48: 1277 (1933).
14. I. A. Baker, A. L. Cochrane, and P. C. Elwood, Letters to the Editor. Growth surveillance Int. J. Epidemiol. 6: 293 (1977).
15. N. Dahlman, and K. Petersen, Influences of environmental conditions during infancy on final body stature, Pediat. Res. 11: 695 (1977).
16. Ministry of Agriculture, Fisheries and Food, Household food consumption and expenditure: 1979 Annual report of the National Food Survey Committee, H. M. S. O., London.
17. A. E. Black, W. Z. Billewicz, and A. M. Thomson, The diets of pre-school children in Newcastle upon Tyne 1968-71. Br. J. Nutr. 35: 105 (1976).
18. M. L. Donnet, T. H. Cole, T. M. Scott, and J. P. Stanfield, Diet, growth, and health in infants in a disadvantaged inner city environment in Glasgow. In "Nutrition and Health". M. R. Turner, ed. M. T. P. Press, Lancaster (1982).
19. J. M. Tanner, and P. B. Eveleth, Urbanization and growth pp 144-166. In "Man in urban environments". G. A. Harrison and J. B. Gibson, eds. University Press, Oxford (1976).
20. R. M. Malina, Secular changes in size and maturity: causes and effects. pp 59-102. In "Monographs of the society for research in child development serial No. 179, 44: Secular trends in human growth, maturation and development", A. F. Roche, ed. The University of Chicago Press, Chicago (1979).
21. N. Cameron, The growth of London schoolchildren 1904-1966. An analysis of secular trend and inner-county variation, Ann. Hum. Biol. 6: 505 (1979).
22. J. M. Tanner, " A history of the study of human growth". Cambridge University Press, Cambridge (1981).
23. F. Galton, On the height and weight of boys 14 years in town and country public schools, J. Anthropol. Inst. 5: 174 (1875-76).
24. A. W. Tuxford, and R. A. Glegg, The average height and weight of English schoolchildren. Br. Med. J. i: 1423 (1911).

25. N. D. Paton, and L. Findlay, Poverty, nutrition and growth. Studies of child life in cities and rural district of Scotland. In "Medical Research Council. Special Report. Series No. 101", H. M. S. O, London (1926).
26. J. W. B. Douglas, and J. M. Blomfield, "Children under five". Allen and Unwin, London (1958).
27. J. M. Foster, S. Chinn, and R. J. Rona, The relation of the height of primary school children to population density, Int. J. Epidemiol. (In Press).

AUXOLOGICAL EPIDEMIOLOGY AND THE DETERMINATION OF THE EFFECTS OF

NOISE ON HEALTH

Lawrence M. Schell

Department of Anthropology
State University of New York at Albany
1400 Washington Avenue
Albany, New York 12222
U.S.A.

Many scientists recognize that the pattern of human physical growth and development constitutes a sensitive response to environmental conditions (Colucci et al., 1973; Johnston and Schell, 1979; Tanner, 1978). Humanists may approach the phenomena of growth and development differently, emphasizing how widespread among the world's societies and cultures is the high value placed on the health of children. The scientific and humanistic approaches may be unified when physical growth and development is taken as a primary measure of community health and of the quality of the environment for human habitation. This use of auxological data and research may be called a type of auxological epidemiology.

Tanner (1979) has thoroughly reviewed the history of auxological epidemiology in Europe and the United States. There the roots of auxological epidemiology are well over 100 years old. The fundamental assumption of contemporary auxological epidemiology is clearly stated by Ferro-Luzzi (this volume): "smaller body size is the response to adverse environmental conditions" and (auxological epidemiology is an approach which) "regards stunting as a waste of human resources and an indicator of environmental pressures." This view is shared by many researchers concerned with public health (Colucci et al., 1973; Johnston, 1981; Knipschild et al., 1981; Schell, 1981; Tanner, 1978).

This approach, however, does not fully agree with the view that growth and development, including reduced growth, is one strategy for human adaptation. In 1969 Lasker developed a paradigm for the study of human adaptation based in part on several investigations in which human growth appeared as an adaptive response to environmental stresses. Physical growth and development, when viewed as development plasticity, may be placed within the context of adaptation along with natural selection, acclimatization, and behavioral adjustments (Lasker, 1969; Stini, 1975a). Lasker's paradigm is an attractive and useful research tool because it directs our scientific attention to adaptive responses that are neither reversible physiological adjustments to maintain homeostasis, nor genetically determined responses to environmental stimuli. Examples of adaptive developmental plasticity are patterns of growth and development which appear to minimize such environmental stresses as hypoxia (Frisancho et al., 1975) and heat or cold (Eveleth, 1966). Though the particular growth patterns described by Lasker (1969), Stini (1975b), Frisancho (1970), and Eveleth (1966) are not demonstrably related to the genetical measure of adaptive success (e.g., genetic fitness), in each case reduced growth may be interpreted as part of a strategy of human biological adaptation. In such an interpretation reduced growth is the outcome of a successful adaptation rather than a sign of poor community health.

The conflict between interpretations of reduced growth becomes sharper because health and adaptation may be interrelated. DuBos brings the two together closely (1965:xvii), "states of health or disease are the expressions of the success or failure experienced by the organism in its efforts to respond adaptively to environmental challenges." Thus reduced growth may signify in the adaptability view successful adaptation (at least in some cases) implying good health, or in the auxological view, poor health implying poor adaptation. The fundamental assumption of auxological epidemiology may need to be tested, or the adaptive significance of reduced growth in the individual or for the population may need to be linked more clearly to significant measures of fitness. Until the interpretation of reduced growth is placed on a firm empirical basis, auxological epidemiologists should be aware of the long term health consequences of reduced growth and of its potential as an adaptive strategy.

Empirical support for the view taken by the auxological epidemiologists comes in part first from laboratory and then epidimiological research on stress. Early in Selye's experiments on stress he noted that many kinds of stress reduced growth in experimental animals (1950). So universally did growth retardation occur in stressed animals that he included reduced growth in the formal description of the General Adaptation Syndrome (G.A.S.). One consequence of this feature of the G.A.S. is that growth deficiency in a population may be used to signal the presence of unfelt but noxious environmental conditions. This application outside the laboratory can be considered a test of the fundamental assumption of auxological epidemiology.

There are a growing number of stresses which we know exert an effect on human growth. Nutrition, infection, high altitude hypoxia, cigarette smoking, and the social conditions that relate to these have been mentioned (Ferro-Luzzi, this volume). To these important and well-researched influences I would add several that may be influential but which require further investigation to establish their potency. Certain chemicals, particularly polychlorinated biphenyls (Yoshimura and Ikeda, 1978) and polybromated biphenyls (Allen et al., 1978, cited in Poland and Cohen, 1980) have been shown to influence growth. The human exposures so far experienced have been acute ones as in the yusho poisoning in Japan in 1968 (Fijisawa and Fujiwara, 1972). Thus, we do not know the effects of chronic exposure to these substances. Nevertheless, the stability and the permanence of these substances in the environment, and the way in which they are distributed in the human food chain suggests that humans may come into contact with them if they have not done so already. Polybromated biphenyls, for example, are more concentrated in the body's fat than in other tissues, and are released in breast milk. A survey of the contents of breast milk from mothers in the southern portion of the U.S. state of Michigan found that about one half of the women have PBB levels in excess of o.07 ppm. while some Michigan women have levels above 20.0 ppm. (Poland and Cohen, 1980). These rather high levels found in some Michigan women are probably traceable to an accident in the manufacture of cattle feed. Manufacturing accidents may be a common source of environmental contamination. As a result of manufacturing, mercury was introduced into the environment of Minamata Bay in Japan. From studies of acute methyl mercury exposure, as in Minamata Bay, we know that mercury produces severe effects on humans, particularly on the developing fetus (Harada and Smith, 1975). Radiation has a detrimental effect on growth as has been documented in the Japanese children exposed to atomic bomb blasts (Burrow et al., 1965), and the Pacific islands children exposed to lower levels of radiation from fallout after atomic bomb testing (Sutow et al., 1965). It is unknown whether smaller effects can be produced by less than such acute exposures.

In addition to these environmental challenges are some environmental features that originate in the increasing urbanization of the human population. The trend towards urbanization which began some eight thousand years ago, is now occurring at a fantastic rate (Davis, 1965). In the experience of urbanization there must be new adjustments, many of which are unnoticed by residents. One feature of the urban environment is air polluted from combustion products, particularly automobile exhaust. Some research has recently been conducted that identified an association between the extent of air pollution where a pregnant woman resided, and the birth weight of her baby (Williams et al., 1977). Further research is necessary to determine the extent and strength of this relationship. Noise is another feature of the urban environment. In most urban areas noise levels are higher than ever before, and more areas have high noise levels than before. Noise has been shown to affect several physiological parameters and the growth of experimental animals (for review see Schell and Lieberman, 1981; Welch and Welch 1970). In general, the urban environment brings children into contact with many new substances, or with familiar ones at higher than usual levels (e.g., lead, noise). This may not have a beneficial effect on child health. As further urbanization occurs and urban-rural differences in growth lessen, researchers might consider the possibility that while the rural environment is improving, the urban environment is changing also.

In the use of auxological data to assess epidemiologically whether a particular, putatively noxious feature of the urban environment is, in fact harmful to health, it is necessary to observe clearly the effects of the variable of interest on growth apart from the influences of other social or biological factors. Since auxological epidemiology takes place outside the carefully controlled laboratory setting, confounding variables cannot be manipulated and excluded. In the natural experiment the social influences on growth must be separated from the effects of the variable under investigation.

There are at least three methods of handling potentially confounding variables. Careful site selection is the primary method. Potentially confounding variables should vary little, or vary randomly with regard to the independent variable under scrutiny. If a completely homogeneous community cannot be found to test the effect of a putatively noxious feature, it may be possible to stratify the sample. Though effective, stratification requires a large sample, especially if several confounding factors are present. In such a case, multivariate statistical techniques may correct for the influence of several extraneous independent variables before testing for a relationship between the independent variable of interest and the dependent growth parameters.

All three of these techniques were employed to investigate the effect of noise upon human prenatal growth. The sample was drawn from a suburban community located adjacent to a large international airport in the U.S. Preliminary inquiries suggested that the community is homogenous in many respects. In personal interviews with nearly 150 families, detailed data were collected on social and biological characteristics (parents' birthplaces, occupations and educational levels, family income and ethnic background, parents' reported heights and weights (which bear a close resemblance to measured height and weight according to Stewart et al., 1980), mother's previous pregnancies and her residences while pregnant). The sample was divided into three groups: high (above the 99 dBA jetplane takeoff contour described by Bragdon, 1971), medium (between the 99 and 90 dBA contours), and low (below the

90 dBA contour). A detailed description of noise levels in this community and the method of determining aircraft noise exposure for individuals requires a lengthy presentation and has been made elsewhere (Schell, 1980a, 1981).

The impression of social homogeneity was born out when the distributions of ethnic backgrounds, mothers' and fathers' occupations, and mothers' and fathers' birthplaces among the three zones were tested by the chi square distribution. All of the distributions appeared random, the largest chi square has an associated probability of only o.33 even though the correction for continuity was not computed, thereby increasing the likelihood of rejection. To further increase the homogeneity of the sample, it was stratified according to the percentage of pregnancy spent outside the area where noise levels had been measured. Excluded from the analysis were births to mothers who had spent more than 30 days of their pregnancy outside the community where noise levels were known. Stratification on this one variable reduced the sample size from 221 to 121 subjects. It is indeed fortunate that the study site is homogenous with respect to the discontinuous variables of parents' occupations, birthplaces and ethnic backgrounds. Had these variables differed between exposure groups, additional stratification would have been required but would have reduced the sample size to nil. Continuous traits such as parents' heights and weights, educational levels and income were studied through correlation analysis. Such independent variables that are associated with the dependent growth variables and whose means differ between exposure groups were considered in multivariate analyses.

Unadjusted means are presented in Table 1. Means of low and medium exposure groups were quite similar (Schell 1980b). When low and medium groups were combined, and then contrasted with the high exposure group, female birth weights differed by 304 grams, a significant difference ($t= -3.65$, $p= 0.01$, with unpooled variances). Male birth weights differed by 180 grams (a nonsignificant difference). Gestation lengths of the most noise-exposed males and females averaged approximately nine days less than the means for the combined low and medium group. This difference did not quite reach statistical significance ($0.05<p<0.10$). An analysis of covariance adjusted the groups' means for the significant associations between growth variables and the continuous biological and social variables that differed among exposure zones. The ANCOVA did not substantially change the pattern of significant results, expect that after covariance adjustment, differences in male gestation length means reached statistical significance.

These results are based on small samples, especially of the most noise-exposed subjects. They must be considered preliminary and viewed with caution. Nevertheless, these results are in agreement with research by other investigators which found a negative

Table 1. Means for high, medium and low noise exposure groups.

	HIGH 100 dBA or more	MEDIUM between 90 and 99 dBA	LOW less than 90 dBA
BIRTH WEIGHT			
Males			
$\bar{X}$	3377	3477	3554
s.d.	674.2	516.8	481.0
n	16	36	15
Females			
$\bar{X}$	3129	3466	3399
s.d.	172.1	435.7	491.6
n	8	32	16
GESTATION AGE			
Males			
$\bar{X}$	270	277	278
s.d.	23.8	11.0	6.1
n	16	37	15
Females			
$\bar{X}$	269	279	279
s.d.	22.4	10.3	14.4
n	8	32	16

effect of noise on growth. In one of the earliest studies on the effects of noise on physical growth, Takahasi and Kyo (1968) found that the frequency of premature births in subjects living closest to a large jet airport was higher than the frequency found further away. Ando and Hattori (1977) studied gravidae from near Osaka International Airport in Japan. They observed human placental lactogen levels (hpl) and related these to the subsequent birth weights. Nearly half of the mothers living close to the airport experienced a drop in hpl levels by the 32nd week of pregnancy. The decrease was equivalent to at least one standard deviation below the mean for the researchers' reference group. The frequency of low birth weight was significantly greater among the group from close to the airport, and there was a significant association between low levels of hpl and low birth weight (Ando and Hattori, 1977). A longitudinal study by the same researchers (1973) of prematurity rates in communities at varying distances from another major airport revealed that rates were similar in the years prior to the introduction of jet plane traffic, but that prematurity in the airport community was increased after jet traffic was begun.

After jet traffic became common, the frequency of birth weights below 2499 grams and over 3000 grams differed significantly between the airport community and the unexposed communities (Ando and Hattori, 1973).

These researchers further suggest that births to mothers residing where noise levels exceed 85 dBA (ECPNL) contributed substantially to the higher prematurity rate. Then, as now, there is no a priori theory to guide researchers who attempt to model the effects of chronic noise exposure. The choice of 85 dBA represents an initial attempt to empirically derive a dose-response curve.

The question of how much noise produces a significant decrease in birth weights was also addressed by Knipschild, et al. (1981). They analyzed weights of nearly 1000 births from the vicinity of Amsterdam Airport. Two analyses were performed, one with the sample divided into low and high noise areas (L_{dn}=60-65 dBA), and then of the high noise sample alone, divided into higher and highest noise areas (L_{dn}=70 dBA). In comparing births from low and high noise areas, they found that after adjusting for confounding factors, birth weights below 3000 grams were more common in the high noise area. The influence of the severest noise exposure was seen in their comparison of the higher and highest exposure groups. Of the 41 births from the highest exposure groups 29% were below 3000 grams. Of 457 births from the high exposure group, 23% were below 3000 grams. Both percentages contrast with the rate of 18% from the low exposure group. The difference in rates was due mainly to differences among females. It is not known why female births should show an effect more than males, indeed the opposite might be expected given the sensitivity of males to environmental insult (Tanner, 1978). Though unusual, Knipschild's findings agree with the results I have presented which show an effect on female birth weights but not on male birth weights. Perhaps the influence on confounding factors is stronger on males, making it more difficult to observe a relatively weak influence such as noise.

In general, the studies reviewed suggest that severe noise exposure may have a detrimental effect upon prenatal development. In addition, each study used a different measure of noise exposure, and a different definition of severe exposure. This reflects the preliminary nature of research on the physiological effects of noise. In the preliminary stage of research on the relationship between growth and almost any environmental feature, the shape of the dose-response curve is unknown. As research is performed, the shape may emerge and it may be learned that some extents of the curve are relatively flat, and no relationship appears there. Had the first researchers probed these flat areas, they might have concluded that no relationship exists between growth and the environmental feature. Before reaching such a conclusion, many types of exposure should be investigated. This process is currently occurring with respect to the influence of alcohol on human prenatal development.

Based on the large body of research on stress, it would seem likely that noise exposure, at some level, would exert an influence on human growth. Laboratory studies of experimental animals exposed to a wide variety of stressors has shown that reduced physical growth is a common stress response (Selye, 1950). Noise is but one of many stimuli (e.g., shock, trauma, cold) capable of eliciting the stress response, and has been shown experimentally to reduce growth in animals (Ames, 1978, Fell et al., 1976, Iturrian and Fink, 1968, Welch and Welch, 1970). It is not known how the growth disturbance is produced. Many laboratory studies have shown that noise produces increased catecholamine, cholesterol and glucocorticoid levels (see Cantrell, 1979 or Schell and Lieberman, 1981, for review) and these changes may influence growth. In many, but not all studies of humans exposed to noise, similar types of changes are produced even at rather low levels of noise (Brandenberger et al., 1977, Cantrell, 1974, Favino et al., 1973, Ortiz et al., 1974, Verdun et al., 1976). Thus, a growth disturbance in humans may be the result of alterations in levels of neurotransmitters or endocrines.

There may be several other pathways by which noise may exert an effect upon growth. Noise influences many systems beyond the endocrine system. A frequently observed response to noise is a change in the cardiovascular system (Andren et al., 1980, Büttig et al., 1980, Borg, 1979, Moskov and Ettema, 1977). Studies of the occupational health of noise-exposed workers suggested that the gastrointestinal tract responds to nervous system stimuli from noise stress (Cohen, 1968, Karagodina et al., 1969). Changes in these systems need to be included in a model of the influences of noise on growth. Finally, noise may produce indirect effects by suppressing appetite or by stimulating behaviors antagonistic to health, especially to prenatal health and growth. Two such behaviors in pregnant women would be cigarette smoking and alcohol consumption. Certainly, before conducting detailed studies to determine the exact biologic pathways linking noise exposure to growth, it is necessary to first observe an association between growth and some level of noise exposure. Such observations are only now beginning to accumulate from various studies. Careful attention must be paid to definitions of severe noise exposure, in order that a description of the dose-response relationship is obtained eventually.

In summary, auxological epidemiology is based on the assertion that reduced growth is a response to adverse environmental conditions, and is indicative of poor health. Some researchers use the sensitivity of growth to determine if a questionable environment is, in fact, unhealthful. When specific environmental stimuli, such as noise, are to be tested, it is necessary to understand the influence on growth of other factors, both social and biologic ones, and to consider their effects on the relationship under investigation. Carefully controlled investigations into the sensitivity of growth to environmental stresses constitute tests of the fundamental assumption of auxological epidemiology and may place our interpretation of the significance of reduced growth on an empirical basis.

REFERENCES

Ames, D. R., 1978, Physiological responses to auditory stimuli, in: "Effects of Noise on Wildlife," J. L. Fletcher and R. G. Busnel, eds., Academic Press, New York.

Ando, Y., and Hattori, H., 1973, Statistical studies on the effects of intense noise during human fetal life, J. Sound and Vibration, 27:101.

Ando, Y., and Hattori, H., 1977, Effects of noise on human placental lactogen (hpl) levels in maternal plasma, Br. J. Obst. Gyn., 84:115.

Andren, L., Hansson, L., Björkman, M., Jonsson, A., and Borg, K., 1978, Hemodynamic and hormonal changes induced by noise, Acta Med. Scand., Suppl., 625:13.

Büttig, K., Zeier, H., Muller, R., and Buzzi, R., 1980, A field study of the vegetative effects of aircraft noise, Arch. Environ. Health, 35:228.

Borg, E., 1979, Physiological aspects of the effects of sound on man and animals. Acta Otolaryngol., Suppl., 360:80.

Bragdon, C., 1971, "Noise Pollution," University of Pennsylvania, Philadelphia.

Brandenberger, G., Follenius, M., and Tremoliere, C., 1977, Failure of noise exposure to modify temporal patterns of plasma cortisol in man, Europ. J. Appl. Physiol., 36:239.

Burrow, G. N., Hamilton, H. B., and Hrubec, A., 1965, Study of adolescents exposed in utero to the atomic bomb, Nagasaki, Japan, JAMA, 192:97.

Cantrell, R. W., 1974, Prolonged exposure to intermittent noise: audiometric, biochemical, motor, psychological and sleep effects, Laryngoscope 84 (10, part 2), Suppl. 1.

Colucci, A. V., Hammer, D. I., Williams, M. E., Hinners, T. A., Pinkerton, C., Kent, J. L., and Love, G. J., 1973, Pollutant burdens and biological response. Arch. Environ. Health, 27:151.

Cohen, A., 1968, Noise effects on health, productivity and well-being, Trans. N. Y. Acad. Sci., 30:910.

Davis, K., 1965, The urbanization of the human population, Sci,

Amer., 213:40.
DuBos, R., 1965, "Man Adapting," Yale University Press, New Haven.
Eveleth, P., 1966, The effect of climate upon growth, Ann. N. Y. Acad. Sci., 134:750.
Favino, A., Maugeri, U., Kauchtschischvilli, G., Robustelli Della Cuna, G., and Nappi, G., 1973, Radioimmunoassay measurements of serum cortisol, thyroxine, growth hormone and lutenizing hormone with simultaneous electroencephalic changes during continuous noise in man, J. Nucl. Biol. Med., 17:119.
Fell, R. D., Ellis, C. J., and Griffith, D. R., 1976, Thyroid responses to acoustic stimulations, Environ. Res., 12:208.
Frisancho, A. R., 1970, Developmental response to high altitude hypoxia, Am. J. Phys. Anthropol., 32:401.
Frisancho, A. R., Velasquez, T., and Sanches, J., 1975, Possible adaptive significance of small body size in the attainment of aerobic capacity among high-altitude Quechua natives, in: "Biosocial Interrelations in Population Adaptation," E. Watts, F. E. Johnston, and G. W. Lasker, eds., Mouton, The Hague.
Fujisawa, H., and Fujiwara, B., 1972, On the influence which PCB exercises on the development of the child, Bull. Fac. Liberal Arts, Nagasaki Univ., Natur, Sci., 13:15.
Harada, M. and Smith, A. M., 1975, Minamata disease: a medical report, in: "Minamata," W. E. Smith, ed., Holt, Rinehart and Winston, New York.
Iturrian, W. B., and Fink, G. B., 1968, Effect of noise in the animal house on seizure susceptibility and growth of mice, Lab. Anim. Care, 18:557.
Johnston, F. E., 1981, Physical growth and development and nutritional status, Federation Proc., 40:2583.
Johnston, F. E., and Schell, L. M., 1979, Anthropometric variation of Native American children and adults, in: "The First Americans: Origins, Affinities, and Adaptations," Gustav Fischer, New York.
Karagodina, I. L., Soldatkina, S. A., Vinokur, I. L., and Klimukhin, A. A., 1969, Effect of aircraft noise on the population near airports, Hyg. Sanit., 34:182.
Knipschild, P., Meijer, H., and Sallé, H., 1981, Aircraft noise and birth weight, Int. Arch. Occup. Environ. Health, 48:131.
Lasker, G., 1969, Human biological adaptability, Science, 166:1480.
Mosskov, J. I., and Ettema, J. H., 1977, Extra-auditory effects in long-term exposure to aircraft and traffic noise, Int. Arch. Occup. Environ. Health, 40:177.
Ortiz, G. A., Argüelles, A. E., Crespin, H. A., Sposari, G., and Villafane, C. T., 1974, Modifications of epinephrine, norepinephrine, blood lipid fractions and the cardiovascular system produced by noise in an industrial medium, Hormone Res., 5:57.
Poland, R. L., and Cohen, S. N., 1980, The contamination of the food chain in Michigan with PBB: the breast feeding question,

in: "Drugs and Chemical Risks to the Fetus and Newborn," R. H. Schwarz, and S. J. Yaffe, eds., A. R. Liss, New York.
Schell, L. M., 1980a, "Environmental Noise and the Prenatal and Postnatal Growth of Children," Ph.D. Dissertation, University Microfilms, Ann Arbor.
Schell, L. M., 1980b, The effect of airport noise on human prenatal growth, paper presented at the 100th meeting of the Acoustical Society of America, Los Angeles, California, November 17-21.
Schell, L. M., 1981. Environmental noise and human prenatal growth, Am. J. Phys. Anthropol., 50:63.
Schell, L. M., and Lieberman, L. S., 1981, Noise stress and cancer, in: "Stress and Cancer," K. Bammer and B. N. Newbery, eds., Hogrefe, Toronto.
Selye, H., 1950, "The Physiology and Pathology of Exposure to Stress," Acta Medical Publishers, Montreal.
Stewart, A. L., Brook, R. H., and Kane, R. L., 1980, "Conceptualization and measurement of health habits for adults in the health insurance study: Vol. II, Overweight," Rand, Santa Monica.
Stini, W. A., 1975a, "Ecology and Human Adaptation," W. C. Brown, Dubuque.
Stini, W. A., 1975b, Adaptive strategies of human populations under nutritional stress, in: "Biosocial Interrelations in Population Adaptation," E. S. Watts, F. E. Johnston, and G. W. Lasker, eds., Mouton, The Hague.
Sutow, W. W., Conrad, R. A., and Griffith, K. M., 1965, Growth status of children exposed to fallout radiation on Marshall Islands, Pediatrics, 36:721.
Takahashi, I., and Kyo, S., 1968, Studies on the differences in adaptabilities to the noisy environment in sexes and in growing processes, J. Anthrop. Soc. Nippon, 76:34.
Tanner, J. M., 1978, "Foetus Into Man," Harvard University Press, Cambridge.
Tanner, J. M., 1979, A concise history of growth studies from Buffon to Boas, in: "Human Growth, Vol. 3, Neurobiology and Nutrition," F. Falkner, and J. M. Tanner, eds., Plenum, New York.
Verdun de Cantogno, L., Dallerba, R., Teagno, P. S., and Cocola, L., 1976, Urban traffic noise, cardiocirculatory activity and coronary risk factors, Acta Otolarygnol., Suppl., 339:55.
Welch, B. L., and Welch, A. M., 1970, "Physiological Effects of Noise," Plenum, New York.
Williams, L., Spence, M. A., and Tideman, S. C., 1977, Implications of the observed effect of air pollution on birth weight, Soc. Biol., 24:1.
Yoshimura, T., and Ikeda, M., 1978, Growth of school children with Polychlorinated Biphenyl poisonong or Yusho, Environ. Res., 17:416.

EFFECTS OF INDUSTRIAL POLLUTION ON SOMATIC AND NEUROPSYCHOLOGICAL DEVELOPMENT

Roland Hauspie*, Marie-Christine Lauwers**, and Charles Susanne

Laboratory of Anthropogenetics
Free University of Brussels
Pleinlaan 2, 1050 Brussels, Belgium

INTRODUCTION

Environmental pollution has become a plague in modern society. The environmental deterioration brought about by industrialization of our society over the past century has been accentuated more recently by a sort of "chemical revolution". There are about 60,000 chemicals in widespread use and about 1000 new substances are added each year (Weinstein, 1981) and the past generations have experienced major exposure to many of these contaminants (Guthrie, 1980). In general, we can state that the environment can become polluted by the release of contaminants in the air, water and soil. Air and water pollution are of prime concern to man due to the more direct contact with these parts of the biosphere. Pollutants in the upper layers of the soil only become a problem when they are biologically transportable to man.

Industrial processes may locally result in poisoning of important parts of the mass of drinking water. Emission of particulates and toxic gases by industries contributes about 15% to the total amount of air pollution. However, the relative health hazard of these industrial pollutants is estimated to be as high as 26% (Murdoch, 1975; Guthrie, 1980). The effects of

---

* National Fund for Scientific Research (Belgium)
** Grant-holder of the Institute for the encouragement of Scientific Research in Industry and Agriculture (I.W.O.N.L.-I.R.S.I.A.)

industrial pollution on man are only poorly understood because
of the limited knowledge of the identity and quantity of
industrial emissions (Guthrie and Perry, 1980). They have mostly
been analyzed in cases of acute intoxication and were generally
considered only to be a problem of occupational health.

A major problem is the chronic effect of adverse environmental
conditions on health in general. Indeed, long-term exposure of a
population to pollutants may result in chronic subclinical
intoxication of its individuals affecting their general health
status without producing direct symptoms of poisoning. This is
denoted by some authors as the non-disease effect (Dolgner et al.,
1974). The impact of this sort of subtle impairment is of major
importance, since large parts of the populations might be exposed
before its dangers are unmasked (Weiss and Spyker, 1974).

Although the harmful effects of pollution on mankind are not
yet fully understood, it is already quite clear that children are
particularly at risk. Some authors suggest that a developing
organism is more sensitive to stressors (Krichagin, 1978) and toxic
effects or that a child shows a more ready absorption of inhaled
or ingested substances, when compared to adults (Darrow and
Schroeder, 1974). Specific behavioral patterns during childhood
or adolescence may also contribute to young people's higher risk
of environmental exposure. The exploratory oral activity of
infants, for example, is a major factor in the problem of lead
intoxication through the ingestion of lead-containing dust and
paint-scales (Charney et al., 1980). Cohen (1974) suggested that
rapid changes in physiologic processes and psychosocial behavior
also make the adolescent unusually vulnerable to certain environmental chemical exposures. He thereby emphasizes the frequently
unique environmental subculture of adolescents, favoring contact
with various drugs, noise and pesticides. Kaufman and Wiese
(1978), for example, showed that gasoline sniffing can be a major
source of lead absorption in children.

A better knowledge of the impact of environmental pollutants
on the health of a population is important since at long term
these factors will adversely affect the rates of morbidity and
mortality as examplified by Holland (1974) and Williams et al.
(1977). Long-term effects can be expected in terms of mutagenicity
and cancerogenicity (Gerber et al., 1980). Physical growth and
development has been considered as a sensitive indicator of the
health of a population (Goldstein and Tanner, 1980; Schmidt and
Dolgner, 1977). Growth retardation proved also to be a more
sensitive measure of environmental hazards than malformations with
regard to the assessment of environmental effects on embryos and
fetuses of experimental animals (Brent, 1974). The interval of
susceptibility to growth retardation by environmental factors is
greater than the interval of susceptibility to teratogenesis.

The aim of the present paper is to review some of the possible effects of industrial pollution on growth and development and to discuss certain interfering factors.

EFFECTS OF TOTAL AIR POLLUTION ON GROWTH AND DEVELOPMENT

Air pollution is caused by a great variety of waste products discharged in the atmosphere. Its main components consist of a number of toxic gases such as carbon monoxide, nitrogen dioxide, sulfur dioxide, hydrogen sulfide and fluorides and of very small particles such as dust, smoke and soot. Some authors (Bachrach et al., 1979) suggested that endogenous vitamin D synthesis was limited by air pollution in densely populated cities and that this factor could contribute to growth retardation. Pelech (1970), in an analysis of 330 prepubertal children from urban and rural environments, observed a retardation of bone age, which could be related to environmental pollution after allowing for socio-economic factors. Confirming results were obtained by Schmidt and Dolgner (1977), who analyzed more than 10,000 children aged 7 to 12 and living in polluted or clean areas. The authors suggested that the delayed skeletal maturation observed in relation to various degrees of air pollution is very likely caused by a lack of ultraviolet radiation. Holland (1974) mentioned the effect of heavy doses of air pollution on growth in Eastern European countries. The author described that the ventilatory function of children aged 5, 11 or 14-16 years is influenced by the degree of pollution, but also by social class and a history of respiratory diseases in early childhood. Air pollution was clearly not the only factor affecting the children.

The implications of air pollution on birth weight have been clearly demonstrated by Williams et al. (1977) in a sample of 424 babies of mothers living in the Los Angeles Basin. The area was characterized by elevated levels of carbon monoxide, nitrogen dioxide and ozone. The authors showed that the degree of air pollution was significantly negatively associated with birth weight after allowing for the effects of other variables which were also significantly associated with birth weight (parity, prepregnant weight, weight gain during pregnancy, number of cigarettes per day, sex and gestation length of the infant). The sample was homogeneous with respect to the socio-economic background so that the authors could not adequately assess a possible interrelation between socio-economic status and pollution level. Average birth weight was reduced by 314 grams in polluted areas when compared to the clear zones, and assuming a linear relationship between birth weight and mortality, the authors calculated that the air pollution could be responsible for a 17% increase in infant mortality.

One of the best documented studies about the effects of air pollution on the growth and development of children was recently published by Thielebeule et al. (1980). They analyzed the evolution of height, weight and bone age over a 10-year period in two different areas of East Germany. One area (Berlin - Pankow) had shown a fairly constant favorable quality of the air between 1968 and 1978, while the other region (Bitterfeld, a center of chemical industry) underwent a substantial improvement of environmental conditions over the same period. This was illustrated by a 50% drop of dust emission. The two samples consisted of about one hundred and eighty 10-year-old children each, both sexes almost equally represented. The change in the degree of air pollution was accompanied by a more favorable development of height, weight and bone age, while there were no appreciable changes of these parameters in the control group. In the polluted area, bone age progressed from about 10 months delay to 6 months (towards chronological age) over the 10-year period of environmental improvement and approached the average bone age for the control group which showed about 4 months delay when compared to chronological age in 1968 and 1978. Finally, Holub and Pelech (1980) failed to find a relationship between growth parameters and varying degrees of air pollution in studying 179 boys of school age. The authors state that the differences in the degree of atmospheric pollution may not have been high enough to trigger differences in the developmental status of the exposed children.

EFFECTS OF CHRONIC EXPOSURE TO LEAD

Significance of subclinical lead intoxication

Clinical intoxication by lead, known as plumbism or saturnism, has decreased during the past decades, notwithstanding the increasing widespread use of lead in various industries (Gerber et al., 1980). This evolution can mainly be attributed to a general improvement of the working conditions and to a tightening up of the hygienic regulations in the factories. Although the average atmospheric lead levels have stopped to increase since 1970 in some industrialized countries (Nriagu, 1980), important amounts of lead are still released in the environment and cause the problem of chronic exposure of the population to low doses of lead-containing products. In the U.S.A., the National Institute of Child Health and Human Development (Publication nr. 82-2348, March 1982) reports that as many as 600,000 children under the age of 6 years show evidence of undue lead absorption, with blood lead levels of 30 to 69 µg/dl. The consequence is that lead may continue to exert a chronic subclinical intoxication in exposed areas, resulting in an increased rate of morbidity and mortality, without producing the classical symptoms of plumbism (Waldron and Stöfen, 1974). Lorenzo et al. (1978) mentioned that only about 0.5% of the children exposed to lead each year exhibit overt clinical symptoms.

Children are particularly at risk for lead intoxication. It was shown by Darrow and Schroeder (1973) that 1 to 5% of urban adults have levels of blood lead above 40 µg Pb/100 ml whole blood, whereas about 25% of the urban children have blood leads at this level. He also reported that 85% of the cases of acute lead intoxication occur in children between 1 and 3 years old. To compare, the natural blood lead concentration in the absence of pollution was estimated by Patterson (1965) to be 0.25 µg Pb/100 ml (or 0.0025 ppm).

The author states that the higher risk for lead intoxication in children is attributable to mainly three factors:
(1) greater sensitivity of children to the toxic effects of lead
(2) greater rate of absorption of lead
(3) greater risk of exposure to lead-containing substances

Sources of lead pollution and ways of uptake

About 5% of the lead in the atmosphere comes from natural sources (mainly windblown dusts and volcanogenic particles) and about 95% from anthropogenic sources. The main contributors of the worldwide lead pollution produced by human activities are: combustion of leaded gasoline (61%), the metallurgic and other industries (23%), mining and smelting of lead (8%) and burning of wood and coal (5%) (Nriagu, 1980). Lead pollution of industrial origin is usually a local problem related to the area around the source of emission. Factors such as rate and point of emission, wind direction and local ventilation are responsible for the daily and seasonal variations in environmental lead concentration. Nriagu also mentioned that the concentration of lead as well as the size of the particulates decrease with distance from the emission source. Contamination by lead can occur through the use of drinking water supplied via lead pipes (Kollath, 1942). Subclinical intoxication by lead through the use of lead-containing drinking water was shown by De Graeve et al. (1975), Rondia and Sartor (1978), Sartor and Rondia (1980) and Sartor and Rondia (1981).

Lead from industrial origin can be taken up by a child through the inhalation of gases, smoke or small particulated substances. Particulates of less than 1 µ reaching the alveola in the lungs are readily absorbed (Lafontaine et al., 1977). Whereas in adults inhalation is the main path of lead absorption, in children lead is most commonly absorbed through ingestion (Jason and Kellogg, 1980). Absorption of lead through the ingestion of lead-containing dust and dust particles is also very important in children because of the relatively greater absorption rate of lead in the intestines of children (50%) (Alexander et al., 1973; Darrow and Schroeder, 1974) compared to adults (6-13%) (Rabinowitz et al., 1976).

Effects of lead on human development

A few studies have been published dealing with the effects of subclinical lead intoxication on the functioning of the brain and the growth and development of children (for a review, see Lauwers, 1981).

Abnormalities in fine-motor control and behavior (negativism, distractibility and constant need for attention) have been observed in 70 children with elevated blood lead levels but with no overt symptoms of lead poisoning (de la Burdé and Choate, 1972). Behavioral disturbance of children (hyperactivity) was also associated with elevated levels of lead in blood (David et al., 1972). More recently, Needleman et al. (1979) gave evidence of reduced intellectual performance in 58 children with elevated dentine lead levels. The frequency of nonadaptive classroom behavior also increased in a dose-related fashion to the amount of lead accumulated in the teeth. The authors concluded that subclinical lead intoxication was responsible for neuropsychologic deficits resulting in reduced classroom performance. The distribution of socio-economic status of the father differed significantly between the exposed and control group but allowance for this was made through an analysis of covariance. The authors failed to show any relationship of lead intoxication with physical development.

Nordström et al. (1978) analyzed birth weight in the offspring of women working in and living around a smelter in Sweden, where several toxic substances, such as arsenic, lead and sulfur dioxide, were emitted. A significantly decreased birth weight was observed and could mainly be attributed to a decrease of birth weight in later pregnancies. Whereas birth weight normally increases with parity, the authors found a decrease in birth weight with parity greater than 2.

One of the first studies indicating impaired postnatal growth and development in children was published by Antal et al. (1968). They compared the psychological and physical development of children in two areas differing greatly in the degree of environmental pollution by lead. The results showed an impaired physical growth and psychological development as well as an increased frequency of adaptation problems and reduced school performances. The study was conducted on about 2300 children as far as the somatic development is concerned and on about 260 subjects for the evaluation of neuropsychological activity. The authors do not give a clear description of the socio-economic background of the sample and control, but state that the differences in the parents' professions between the two groups were small and non-significant.

Finally, Habercam et al. (1974) found a statistically significant negative relationship between blood and tooth lead on the one hand and growth parameters (height and weight) on the other in an analysis of 18 preadolescent children. They calculated that about 22% of the growth deficit could be attributed to lead. No information was given as to the possible interaction of other factors with the observed relationship between lead levels and growth.

Effects of lead on experimental animals

Many studies on experimental animals, mainly on mammals and particularly on rats, have shown that chronic lead poisoning causes a reduction in fertility (Grant et al., 1980), a retardation of fetal and postnatal growth (Baernstein and Grand, 1942; Michaelson, 1973; Michaelson and Sauerhoff, 1974), neurological damage (Pentschew and Garro, 1966; Bornschein et al. 1980) and an increased rate of fetal and postnatal deaths (Gerber et al., 1980). The results of these investigations contribute to a better understanding of problems such as dose-effect relationships, mechanisms of action and influences of other factors on lead toxicity.

Exposure to lead of neonatal rats through the milk of the dams causes immediate and long-term disturbance of the central nervous system functioning (Fox et al., 1979). Golter and Michaelson (1975) observed neurological manifestation of lead poisoning in suckling rats associated with increased levels of norepinephrine, suggesting a possible relationship between lead intoxication and altered behavior (increased motor activity). Ahrens and Vistica (1977) produced lead encephalopathy in suckling rats by exposing lactating mothers to lead in their drinking water immediately after birth. There was evidence that the neurological symptoms were primarely due to microvascular damage of the brain. Hemorrhagic encephalopathy was also observed by Lefauconnier et al. (1980) in suckling rats after exposure to high levels of lead. However, they suggested that subtle dysfunction of the nervous system, induced by low levels of lead, are due to a direct action of lead on brain cells rather than on endothelial cells of the capillaries.

In a recent study, McCauley et al. (in press) showed that prenatal and postnatal lead exposure in rats had an inhibiting effect on synaptogenesis in the cerebral cortex. Other work done in the same laboratory had shown that the same conditions of lead intoxication also induced delays in locomotor and exploratory behavior of the young (Crofton et al., 1980) as well as learning deficits in the young rats (Taylor et al., in press). A major problem in evaluating the results of these studies with a view to an extrapolation to man seems to be the inability to produce neurological changes in some laboratory animals (at least in the adults) comparable to human lead encephalopathy. This would seem to be due to a much greater resistance to nervous system effects of lead poisoning (Bornschein et al., 1977; Pentschew and Garro, 1966). But on the other hand, Lorenzo et al. (1978) think that the effects of lead toxicity on the central nervous system in rabbits closely parallel those in man, an opinion which is not at all shared by Michaelson (1973). Perhaps, one should use only humanlike animals (Gerber et al., 1980) since it was shown that rhesus monkeys, for example, present a pattern of sensitivity towards lead intoxication similar to that of man (Allen et al., 1974). It appears that maternal behavior in rats is not a confounding factor in the investigation of the harmful effects of chronic lead exposure on the subsequent offsprings (Zenick et al., 1979).

The effects of lead poisoning on body weight, and in some cases on body length, of growing laboratory animals was clearly demonstrated in rats by Grant et al. (1980), Hsu (1981), Michaelson (1973), Aviv et al. (1980), Kimmel et al. (1980); in mice by Gerber and Maes (1978); in rabbits by Lorenzo et al. (1978); in chickens by Gilani (1973); in hamsters by Carpenter and Ferm (1977); and in dogs by Penumarthy et al. (1980). Gerber and Maes (1978) suggested that disturbed blood formation in lead intoxicated mouse embryos (through impaired heme synthesis) limits their body growth towards the end of pregnancy. It is assumed that, besides the erythropoeitic system, also the kidney and perhaps the gastro-intestinal system are target organs of lead poisoning. Reduced kidney function in rats as a result of lead intoxication (Aviv et al., 1980; Fowler et al., 1980) is supposed to be related to the retarded body weight (Wapnir et al., 1977). These authors further suggested that growth failure in rats under exposure to lead acetate is not due to decreased food intake but can be attributed to malabsorption of the food in the intestine. However, Michaelson and Sauerhoff (1974) demonstrated in a paired feeding experiment that the growth of young rats suckling lead intoxicated mothers showed a growth retardation which could be related to a nutritional effect rather than a direct effect.

The susceptibility of experimental animals to lead poisoning has proved to be influenced by the nutritional status previous to the onset of exposure. A poor diet preceding the lead intoxication may alter some of the deleterious effects of the poisoning (Wapnir et al., 1980). Chisolm (1980) also emphasized the availability of trace elements, such as calcium, iron, copper and zinc in the diet. Deficient intake of these elements enhanced lead absorption and retention. Although the exact mechanisms of interaction have still to be elucidated from animal experiments it appears that they are also relevant to the human situation since those groups of children at risk for chronic lead intoxication may also show prevalence of deficient dietary intakes of calcium and iron (Chisolm, 1974). It is obvious that a supplement of milk (important source of calcium for man) to the daily diet could in part help to prevent some of the toxic effects of chronic lead poisoning.

REFERENCES

Ahrens, F.H. and Vistica, D.T., 1977, Microvascular effects of lead in the neonatal rat, I. Histochemical and light microscopic studies, Experimental and Molecular Pathology, 26:129.

Alexander, L., Delves, H.T. and Clayton, B.E., 1973, The uptake and excretion by children of lead and other contaminants, in : "Environmental health aspects of lead," D. Barth, A. Berlin and R. Engel, ed., Commission of the European Communities, Center for information and documentation, Luxembourg.

Allen, J.R., Mc Wey, P.J. and Suomi, S.J., 1974, Pathobiological and behavioral effects of lead intoxication in the infant rhesus monkey, Environmental Health Perspectives, 7:239.

Antal, A., Timaru, J., Muncaci, E., Ardevan, E., Ionescu, A. and Sandulache, L., 1968, Les variations de la réactivité de l'organisme et de l'état de santé des enfants en rapport avec la pollution de l'air communal, Atmospheric Environment, 2:383.

Aviv, A., John, E., Bernstein, J., Goldsmith, D.I. and Spitzer, A., 1980, Lead intoxication during development, its late effects on kidney function and blood pressure, Kidney International, 7:430.

Bachrach, S., Fisher, J. and Parks, J.S., 1979, An outbreak of Vitamin D deficiency rickets in a susceptible population, Pediatrics, 64:871.

Baernstein, H.D. and Grand, J.A., 1942, The relation of protein intake to lead poisoning in rats, Journal of Pharmacology and Experimental Therapeutics, 74:18.

Bornschein, R.L., Michaelson, I.A., Fox, D.A. and Loch, R., 1977, Evaluation of animal models used to study effects of lead on neurochemistry and behavior, in : "Biochemical effects of environmental pollutants," S.D. Lee and B. Peirano, ed., Annarbor Science, Michigan, 441-460.

Bornschein, R.L., Pearson, D. and Reiter, L., 1980, Behavioral effects of moderate lead exposure in children and animal models, part 2, animal studies, CRC Critical Reviews in Toxicology, 8:101.

Brent, R.L., 1974, Environmental effects on the embryos and fetuses of experimental animals, Pediatrics, 53:821.

Carpenter, S.J. and Ferm, V.H., 1977, Embryopathic effects of lead in the hamster, a morphologic analysis, Laboratory Investigations, 37:369.

Charney, E., Sayre, J. and Coulter, M., 1980, Increased lead absorption in inner city children: where does the lead come from ?, Pediatrics, 65:226.

Chisolm J.J., 1974, Heavy metal exposure : toxicity from metal-metal interactions and behavioral effects, Pediatrics, 53:841.

Chisolm, J.J., 1980, Lead and other metals : a hypothesis of interactions, in : "Lead toxicity," R.L. Singhal and J.A., Thomas, ed., Urban & Schwarzenberg, Baltimore, 461-482.

Cohen, M.I., 1974, Environmental exposures and biologic considerations of the adolescent in relation to chemical pollutants, Pediatrics, 53:845.

Crofton, K.M., Taylor, D.M., Bull, J.R., Sivulka, D.J., Lutkenhoff, S.D., 1980, Developmental delays in exploration and locomotor activity in male rats exposed to low level lead, Life Sciences, 26:823.

Darrow, D.K. and Schroeder, H.A., 1974, Childhood exposure to environmental lead, Advances in Experimental Medicine, 48:425.

David, O., Clark, J. and Voeller, K., 1972, Lead and hyperactivity, The Lancet, Oct. 28:900.

De Graeve, J., Jamin, P. and Rondia, D. 1975, Plombémie d'une population adulte de l'est de la Belgique, Revue d'épidémiologie, médecine sociale et santé publique, 23:131.

de la Burdé, B. and Choate, M.S., 1972, Does asymptomatic lead exposure in children have latent sequelae ?, Journal of Pediatrics, 81:1088.

Dolgner, R., Pelech, L. and Schmidt, P., 1974, Die Reaktion des Schulkindes auf Luftverunreinigungen-Ergebnisse und Erfahrungen aus mehrjährigen Untersuchungen, Zentralblatt für Bakteriologie, Parasitenkunde, Infektionskrankheiten und Hygiene, I Abt. Orig. A., 227:101.

Fowler, B.A., Kimmel, C.A., Woods, J.S., McConnell, E.E., and Grant, L.D., 1980, Chronic low-level lead toxicity in the rat. III. An integrated assessment of long-term toxicity with special reference to the kidney, Toxicology and Applied Pharmacology, 56:59.

Fox, D.A., Overmann, S.R. and Woolley, D.E., 1979, Neurobehavioral ontogency of neonatally lead-exposed rats II. Maximal electroshock seizures in developing and adult rats, Neurotoxicology, 1:149.

Gerber, G.B., Léonard, A. and Jacquet, P., 1980, Toxicity, mutagenicity and teratogenicity of lead, Mutation Research, 76:115.

Gerber, G.B. and Maes, J., 1978, Heme synthesis in the lead-intoxicated mouse embryo, Toxicology, 9:173.

Gilani, S.H., 1973, Congenital anomalies in lead poisoning, Obstetrics and Gynecology, 41:265.

Goldstein, H. and Tanner, J.M., 1980, Ecological considerations in the creation and the use of child growth standards, The Lancet, 168:582.

Golter, M. and Michaelson, I.A., 1975, Growth, behavior and brain catecholamines in lead-exposed neonatal rats : a reappraisal, Science, 187:359.

Grant, L.D., Kimmel, C.A., West, G.L., Martinez-Vargas, C.M. and Howard, J.L., 1980, Chronic low-level lead toxicity in the rat II. effects on postnatal physical and behavioral development, Toxicology and Applied Pharmacology, 56:42.
Guthrie, F.E., 1980, Nonagriculture pollutants, in: "Introduction to environmental toxicology," F.E. Guthrie and J.J. Perry, ed. Blackwell scientific publications, Oxford, 1-33.
Guthrie, F.E. and Perry, J.J., 1980, Introduction to environmental toxicology, Blackwell scientific publications, Oxford.
Habercam, J.W., Keil, J.E., Reigart, J.R. and Croft, H.W., 1974, Lead content of human blood, hair and decidious teeth: correlation with environmental factors and growth, Journal of Dental Research, 53:1160.
Holland, W.W., 1974, Effects of air pollution on children, Pediatrics, 53:839.
Holub, M. and Pelech, L., 1980, Some growth and hematological indices in children of school age living in different localities of the district Mělník, Československá Hygiena, 25:158.
Hsu, J.M., 1981, Lead toxicity as related to glutathione metabolism, The Journal of Nutrition, 111:26.
Jason, K.M. and Kellogg, C.K., 1980, Behavioral neurotoxicity of lead, in: "Lead toxicity," R.L. Singhal and J.A. Thomas, ed., Urban & Schwarzenberg, Baltimore, 241-271.
Kaufman, A. and Wiese, W., 1978, Gasoline sniffing leading to increased lead absorption in children, Clinical Pediatrics, 17:475.
Kimmel, C., Grant, L.D., Sloan, C.S. and Gladen, B.C., 1980, Chronic low-level lead toxicity in the rat I. maternal toxicity and perinatal effects, Toxicology and Applied Pharmacology, 56:28.
Kollath, W. 1942, Uber die aussergewerbliche Bleiaufnahme durch Trinkwasser, ihre klinische Bedeutung, bevölkerungspolitische Gefahr und ihre Verhütung, Münch. Med. Wschr., 14:927.
Krichagin, V.J., 1978, Health effects of noise exposure, Journal of Sound Vibration, 59:65.
Lafontaine, A., Aerts, J., Bruaux, P., Claeys-Thoreau, F., Impens, R., Mathy, P., Roels, H. and Van Bruwaene, R., 1977, Lood in het leefmilieu in België, Belgisch Archief van Sociale Geneeskunde, Hygiene, Arbeidsgeneeskunde en Gerechterlijke Geneeskunde, 35:451.
Lauwers, M.-C., 1981, De gezondheidskundige betekenis van lood in de omgeving van de mens en de invloed van lood op de ontwikkeling van mens en dier, dissertation Certificate in Human Ecology, Vrije Universiteit Brussel, Brussels.
Lefauconnier, J.M., Lavielle, E., Terrien, N., Bernard, G. and Fournier, E., 1980, Effect of various lead doses on some cerbral capillary functions in the suckling rat, Toxicology and Applied Pharmacology, 55:467.

Lorenzo, A.V., Gewirtz, M. and Averill, D., 1978, CNS lead toxicity in Rabbit offspring, Environmental Research, 17:131.

McCauley, P.T., Bull, R.J., Tonti, A.P., Lutkenhoff, S.D., Meister, M.V., Doerger, J.V. and Stober, J.A., The effects of prenatal and postnatal lead exposure on neonatal synaptogenesis in rat cerebral cortex, Journal of Toxicology and Environmental Health, in press.

Michaelson, I.A., 1973, Effects of inorganic lead on RNA, DNA and protein content in the developing neonatal rat brain, Toxicology and Applied Pharmacology, 26:539.

Michaelson, I.A. and Sauerhoff, M.W., 1974, An improved model of lead-induced brain dysfunction in the suckling rat, Toxicology and Applied Pharmacology, 28:88.

Murdoch, W.W., 1975, Environment. Sunderland, Mass., Sinauer Associates.

National Institute of Child Health and Human Development, Research programs, US Department of Health and Human Services, 1982, Mental retardation and developmental disabilities, Publication n° 82-2348.

Needleman, H.L., Gunnoe, C., Leviton, A., Reed, R., Peresie, H., Maher, C. and Barrett, P., 1979, Deficits in psychologic and classroom performance of children with elevated dentine lead levels, The New England Journal of Medicine, 300:689.

Nordström, S., Beckman, L. and Nordenson, I., 1978, Occupational and environmental risks in and around a smelter in northern Sweden, Hereditas, 88:43.

Nriagu, J.O., 1980, Lead in the atmosphere and its effects on lead in humans, in : "Lead toxicity," R.L. Singhal and J.A. Thomas, ed., Urban & Schwarzenberg, Baltimore, 483-503.

Patterson, C.C., 1965, Contaminated and natural lead environments of man, Archives of Environmental Health, 11:344.

Pelech, L., 1970, Retardation of skeleton age of children from urban and rural areas with increased atmospheric pollution, Fortschr. Med., 88:510.

Pentschew, A. and Garro, F., 1966, Lead encephalo-myelopathy of the suckling rat and its implications on the porphyrinopathic nervous diseases, Acta Neuropathologica, 6:266.

Penumarthy, L., Oehme, F.W. and Galitzer, S.J., 1980, Effects of chronic oral lead administration in young beagle dogs, Journal of Environmental Pathology and Toxicology, 3:465.

Rabinowitz, M.B., Wetherill, G.W. and Kopple, J.D., 1976, Kinetic analysis of lead metabolism in healthy humans, The Journal of Clinical Investigations, 58:260.

Rondia, D. and Sartor, F., 1978, Conséquences pour la santé d'une interaction entre l'eau potable douce et les matériaux, La Tribune du Cebedeau, 419:341.

Sartor, F.A. and Rondia, D., 1980, Blood lead levels and age : a study in two male urban populations not occupational exposed, Archives of Environmental Health, 35:110.

Sartor, F.A. and Rondia, D., 1981, The setting of legislative norms for the environmental lead exposure : results of an epidemiological survey in the east of Belgium, Toxicology Letters, 7:251.
Schmidt, P. and Dolgner, R., 1977, Zur Deutung einiger Befunde aus Kinderuntersuchungen in Gebiete met unterschiedlich starker Luftverunreinigung, Zentralblatt für Bakteriologie, Parasitenkunde, Infektionskrankheiten und Hygiene, I Abt. Orig.B, 165:539.
Taylor, D.H., Noland, E.A., Brubaker, C.M. and Bull, R.J., Low level lead (Pb) exposure produces learning deficits in young rat pups, Neurobehavioral Toxicology, In press.
Thielebeule Von, U., Pelech, L., Grosser, P.J. and Horn, K., 1980, Körperhöhe und Knochenalter bei Schulkindern in lufthygienisch unterschiedlich belasteten Gebieten, Z. Ges. Hyg., 26:771.
Waldron, H.A. and Stöfen, D., 1974, Sub-clinical lead poisoning, Academic Press, London.
Wapnir, R.A., Exeni, R.A., McVicar, M. and Lifshitz, F., 1977, Experimental lead poisoning and intestinal transport of glucose, amino acids and sodium, Pediat. Res., 11:153.
Wapnir, R.A., Moak, S.H. and Lifshitz, F., 1980, Malnutrition during development : effects on later susceptibility to lead poisoning, The American Journal of Clinical Nutrition, 33:1071.
Weinstein, B., 1981, The scientific basis for carcinogen detection and primary cancer prevention, Cancer, 47:1133.
Weiss, B. and Spyker, J.M., 1974, Behavioral implications of prenatal and early postnatal exposure to chemical pollutants, Pediatrics, 53:851.
Williams, L., Spence, M.A. and Tideman, S.C., 1977, Implications of observed effect of air pollution on birth weight, Social Biology, 24:1.
Zenick, H., Pecorraro, F., Kennedy, D.P. and Ward, J., 1979, Maternal behavior during chronic lead exposure and measures of offspring development, Neuro-behavioral Toxicology, 1:65.

DISCUSSION REPORT

(Ecological Environment)

Ajit Ray

The presented papers in "Ecological Environment and Growth" workshop cover a broad spectrum of enquiry revolving around the central theme of Growth and its impairment, in association with an investigation into the underlying causes and influences which are responsible for the observed variation in human populations.

The paper on "Environmental and Physical Growth" by Dr. Ferro-Luzzi, is a critical review of the recent literature on environmental causes of growth retardation in man. Only the issues most relevant to public health and population groups have been selected. Adverse environmental conditions cause growth failure. Under these conditions, small body size is considered the outcome of an adaptive strategy. Although some authors consider stunting beneficial in terms of population, the more sound view considers it wasteful of human resources (associated as it is with permanent impairment of mental development, reduced work capacity, and higher morbidity and mortality rates) and indicative of undesirable environmental pressures. The environmental stressors associated with growth failure are nutritional factors, infectious diseases, high altitude hyposia, socioeconomic conditions, smoking during pregnancy, neurohumoral disturbances secondary to psychological deprivation, and temperature or climate extremes. A nearly infinite combination of duration, severity, timing, and type of stressor is theoretically possible, with each combination stimulating differently the adaptive processes and eliciting a wide range of responses. The timing of the exposure to stress is important. Disturbing factors may interfere either by delaying the maturation process ("timing effect") or by directly preventing the realization of growth potential ("intensity effect"). As a study of chronically malnourished Ladino children (Martorell et al., 1979) has shown, maturation may be delayed to a lesser extent than growth, with different effects on body size. Food supplementation improved both the growth performance and skeletal maturity of the children but at a differential rate.

Weight and height, the conventional indicators of body size, have been universally employed as dependent variables in the predive equations describing the impact of environment on genotype. Other indices occasionally used are: skinfold thickness, upper arm muscles area or circumference, sitting height, and laterality. The direct determinants of growth performance are food intake and infectious diseases. The indirect determinants are all those conditions that affect food intake and facilitate the spread of infections. High altitude hypoxia and smoking during pregnancy represent a special category of indirect determinants.

Professor Schell discusses a number of stresses known to affect human growth and development in his paper on "Auxological Epidemiology and the Determination of the Effects of Noise on Health". The approach taken is that of auxological epidemiology, which equates poor growth with poor health, unlike that of some of the adaptationists, who may view poor growth as a relatively successful adaptation to some environmental conditions.

Empirical data are cited in support of the former view, showing that a common response to stress in laboratory animals exposed to a wide variety of stressors is reduced physical growth. In addition to diverse environmental challenges, as pointed out by Dr. Ferro-Luzzi, is the challenge created by the increasing urbanization of the human population which requires adjustements to such stresses as air and noise pollution. It is suggested that growth deficiency my be used to signal the presence of unfelt, but noxious environmental conditions.

The findings of the present investigation are compared to other studies. The present sample was drawn from a suburbian community adjacent to a large international airport in the U.S.A. and divided into three groups: high noise exposure (average exposure above 99 dBA), medium noise exposure (average exposure between 90 and 99 dBA) and low noise exposure (average exposure below 90 dBA) during pregnancy.

The findings showed that the means of the low and medium exposure groups were quite similar; when low and medium groups were combined and then compared with the high exposure group, female birth weights differed significantly (at $p = 0.01$) by 304 grams, but male birth weights differed only by 180 grams - a nonsignificant difference. The reason for the different results for the sexes was unknown. The gestation period was 8 days less for the high exposure group, a difference that is not statistically significant.

Dr. Hauspie in his paper on "Effects of Industrial Pollution on Somatic and Neuropsychological Development" reviews some of the possible effects of total air pollution and chronic exposure to cadmium and lead on the growth and development of man and experimental animals.

It points out that approximately 1,000 new chemical substances are manufactured annually, and produced on a large scale. The harmful effects of pollution on man are not fully understood because of the limited knowledge of the identity and quantity of industrial emissions, but the chronic effects of adverse environmental conditiond pose a major health problem and long term effects may result in chronic subclinical intoxication which may affect general health status without producing direct symptoms of poisoning. The study stresses the need for a better understanding of the impact of environmental pollutants on population health, since the long term effects will include higher morbidity and mortality rates.

The subclinical effects of chronic exposure to lead are not well known. Laboratory experiments suggest that permanent damage to the central nervous system and severe growth retardation occur. With chronic lead poisoning, there is indication if reduced fertility, retardation of fetal and postnatal growth, neurological damage, and an increased rate of fetal and postnatal mortality. Aculating evidence from studies of children with subclinical levels indicates that both psychological and physical development can be impaired - with recovery possible in the latter, but unlikely in the former. In addition, there is evidence of impairment of fine motor control, behaviour disorders, reduced intellectual performance, reduced birth weights in infants, and impaired postnatal growth and development.

Dr. Rona commenting on Ferro-Luzzi's paper on "Environment and Physical Growth" points out that a major difficulty Ferro-Luzzi faces in discussing a topic as broad as the impact of the environment on physical growth is the absence of a unifying theory explaining all the observed phenomena and at the same time giving guidance for intervention in communities. Rona's comments on Ferro-Luzzi's paper are aimed at discussing the flaws and omissions in research on environment and growth. They are restricted primarily to the social environment and make the community, as distinct from the individual, their centre. Rona als points out that most of his experience in this field has been in the U.K. His paper is divided into three sections: "Social Class and Growth", "Assessing the Impact of Economic Recession and National Disasters on Children's Growth", and "Degree of Urbanization and Growth: An Odd Finding".

The study of human growth and development encompasses the field of human ecology, human population genetics and developmental processes. Populations do not have fixed relations with the environment; they are sensitive to the environment and change their genetic constitutions through interactions with other populations and in response to environmental changes and influences. Growth and developmental processes though, are under genetic control, are influenced by the environment and the resulting phenotype is the product of gene and environment.

In spite of careful collection of data and searching analysis of many interacting social and ecological variables, it was apparent from the discussions, how complex the problems are in the field of Ecological Environment and Human Growth. What should we use as a control group for growth study: general population, sub-population or other specific groups? How do we define socio-economic strata satisfactorily? How can we distinguish growth patterns between overall aspects of defined groups and different trends within any one group? Are small statured people better adapted then tall people? How do we define the concept of Darwinian fitness and adaptedness in growth study?

Dr. Rona has pointed out that most surveys have shown the tendency of children in lower social classes to be shorter than those in non-manual social classes, but there are few longitudinal studies of children to determine more precisely the effects of social factors on growth over a period of time. Those that exist do not always agree. For example, Dr. Ferro-Luzzi quoting a paper by Dr. Garn comments that differences among children from different classes are already apparent at birth and increase progressively up to the age of seven, indicating a cumulative environmental effect. Longitudinal analyses of the relative heights of children from different social classes done in the U.K., however, suggest that the associations between attained height and social factors arise almost entirely before schooling. Furthermore, social class is an example of complex variables. It is a proxy measure of family income and lifestyle, but this distinction of the two components is very rough, since lifestyle is made up of a complex set of behaviours. If an attempt to distinguish between the effects of these two major components of social class were made and they proved mutually exclusive, intervention strategies could be better designed and evaluated.

Dr. Rona also pointed out that Ferro-Luzzi has shown that in Italy, in the past, children in rural areas were smaller than those living in cities and that improvements in socio-economic circumstances since 1950 have eroded this difference. The majority view has been that urbanization brings increased economic benefits, which result in improved nutrition and child care.

In the U.K., the secular trend of growth has been well documented, but, over the last 300 years, English and Scottisch studies show, not that rural children are shorter than urban children, but that an inverse relationship between height and population density exists in the U.K. Thus, at least in the U.K., urbanization _per se_ cannot be the cause of the observed secular trend of growth. Urbanization may have been a confounding variable in the process and may reflect the association of this with general improvement in the standard of living during this century in many other countries. The U.K. results should lead us to be more critical of commonly accepted views of the relationship between growth and environment.

Even for those aspects of the environment that have been more thoroughly researched, there are serious gaps. If greater effort is not made to fill them, auxology will produce only descriptive studies and oversee nutritional status, but it will scarcely contribute to further advance in the social and health care of children in the community.

Adverse environmental conditions, for example, which encompass nutritional factors, socio-economic conditions, infectious diseases, pollution and social and cultural practices, are considered some of the main contributors to growth retardation, morbidity and mortality. For instance, as Dr. Hoang asked during discussion period, what is the effect of increased pollution, increased absence from home, and consequent psychological deprivation on the progeny of working mothers? Do these children suffer obvious consequences such as impairment in growth or a subsequent deterioration in health, or do they "adapt" to the changing environment? These, and similar questions, require the simultaneous examination of multiple and interacting influences. Only then can we extrapolate relevant and meaningful information. For instance, as pointed out by both Professor Schell and Dr. Hauspie, whether one is looking at noise or industrial pollution, it is imperative to know the levels and duration of exposure, which in turn, are governed by location, occupation, socio-economic condition, and the like. To know that a population is at risk of exposure, without knowing concomitant factors, is to know very little. Similarly, to know that adverse environmental conditions can cause growth retardation,is insufficient if one does not know how or why. As pointed out by Dr. Ferro-Luzzi, the environmental stressors associated with growth failure form almost infinite combinations of severity, timing, and duration of types of stressors, with each combination stimulating differently the adaptive processes and eliciting a wide range of responses from the growing individual.

Different populations and different aspects of the same population, vary in their developmental sensitivity to external

variation. Therefore, concepts regarding health measure have to be
clearly understood and defined. For instance, the auxological
epidemiology approach, as presented by Professor Schell, equated
poor growth with poor health, whereas some adaptionists view poor
growth as a relatively successful adaptation to harsh environmental
conditions. There are many measures of health which are dependent
upon such elusive criteria as the values and resources of the
society being studied. For example, if resources are scarce in a
society, one may use mortality as a measure of health. In a society
with more resources, morbidity may be included as an additional
measurement. As shown by Dr. Hauspie, in societies with medical
and social benefits, even subclinical levels of various toxic agents,
can be used as indicators of health. Consequently, equating poor
health with poor growth requires an investigation of the total
sociobiological system.

Dr. Bailey initiated the discussion on the question of nature
of heritability of stature in different populations. The heritability refers to the amount of the total pheneotypic variation attributable to genetic variation, combining together with additive interaction, dominance and assortive mating effects. The heritabilities
of quantitative traits in man vary widely for different traits, and
there is evidence that much of the variation in continuous traits,
like stature, is genetic in origin. However, the heritability of
stature measures the relative importance or heredity in comparison
with environment on the variance of the trait in a specific population at a specific time. In addition, a particular gene may have
different phenotype effects in different environments.

It was pointed out during discussion, that unless we are clear
what we mean by Darwinian fitness and adaptation in growth study,
our reasoning becomes circular. Both Darwinian fitness and adaptation are responses of a population's gene pool to a change in the
environment. Darwinian fitness, however, is not the same as
adaptedness. Adaptedness may be measured in absolute terms, whereas
Darwinian fitness is relative to other genotypes. Darwinian fitness
may be discussed in terms of selective advantages of different
genotypes but it provides very little evidence in which natural
selection takes place at the phenotypic level. Both adaptation
and fitness, however, must be viewed in terms of time and environment. Considering either without the context of environment reduces
the definitions. For instance, the adaptation of body size, shape
and composition to extreme climatic conditions has been well documented over the years. The question of whether small is better
than tall is totally dependent upon where one lived. A body shape
which aids the dissipation of excess heat is a useful adaptation
in hot climates, whereas cold climates dictate physiques which
facilitate the conservation of heat. However, much remains to be
done in understanding the adaptive value of human stature in
Mendelian population.

The participants in this discussion were:

Bailey, Ferro-Luzzi, Hauspie, Hoang, Mueller, Rona, Tanner, Schell, Van Wieringen, Neyzi, and Roche.

CONTRIBUTORS

Brandt, I.	Universitäts-Kinderklinik, Adenauerallee 119, 5300 Bonn, W. Germany
Bock, R.D.	University of Chicago, Chicago, Illinois 60637 USA
Cameron, N.	Dept. of growth and development, Inst. of Child Health, 30 Guilford Street, London WC 1N 1EH, England
Ferro-Luzzi, A.	Inst. Nazionale della Nutrizione, Via Ardeatina 546, 00179 Roma, Italy
Greco, L.	Dept. Paediatrics, Faculty of Medecine, Napoli, Italy
Hauspie, R.	Vrije Universiteit Brussel, Lab. Antropogenetika, Pleinlaan 2, 1050 Brussels, Belgium
Klissouras, V.	Sports Research Institute, 37 Kifissias Ave., Maroussi, Athens, Greece
Lammert, O.	Odense University, Inst. Physical Education, Campus vej 55, DK-5230 Odense M, Denmark
Lauwers, M.C.	Vrije Universiteit Brussel, Lab. Antropogenetika, Pleinlaan 2, 1050 Brussels, Belgium
Massé, G.	Ecole Nationale Santé Publique, Ave. dr. Pr. Léon Bernard, 35043 Rennes Cédex, France
Mayer-Greco, M.	Dept. Paediatrics, Faculty of Medecine, Napoli, Italy
Mueller, W.H.	University of Texas Health Science Center School of Public Health, P.O. Box 20186, Houston, Texas 77225, USA

Neyzi, O.	Dept. Pediatrics, Istanbul Faculty Medicine, Capa, Istanbul, Turkey
Ray, A.	Dept. of Anthropology, Univ. of Toronto, 100 St. George Street, Toronto, Canada
Roede, M.J.	Wilhelmina Kinderziekenhuis, Nieuwe Gracht 137, 3512 Utrecht, Nederland
Rona, R.	Dept. of Community Medicine, St. Thomas Hospital, London SE 1, England
Schell, L.	State University of New York at Albany, Dept. Anthropology, 1400 Washington Avenue, Albany, New York 12222, USA
Smith, J.	Dept. Human Biology and Health, University of Surrey, Guilford, Surrey, England
Susanne, C.	Vrije Universiteit Brussel, Lab. Antropogenetika, Pleinlaan 2, 1050 Brussels, Belgium
Van Wieringen, J.V.	Wilhelmina Kinderziekenhuis, Nieuwe Gracht 137, 3512 Utrecht, Nederland
Watts, E.	Department of Anthropology, Tulane University, New Orleans, Louisiana 70118, USA

INDEX

Adaptation
   developmental plasticity, 169, 210, 235, 240
Additive factors, 63, 66
Age classes, 38, 39
Age range, 38
Air pollution, 211, 221-228 (*See also* Birthweight, Pollution, industrial)
Altitude (high), 51, 149, 169, 180, 235 (*see also* Placenta)
Animals
   experimental, 85, 155, 216, 227, 228, 236, 237
Anthropometry, 91-96
Appropriate for gestational age, 101-103, 105-114, 126, 161 (*see also* Small for gestational age)
Assortative mating, 52, 75, 93, 95
Athletes, 81
Autocorrelation, 34

Behavior
   and lead, 226 (*see also* Intellectual performance, Blood lead level)
Biochemistry, 49
Biparietal diameter, 127, 160
Birthweight, 103, 110, 117, 124-130, 146-151, 162, 170, 175-177, 180, 181, 201, 213-215, 236 (*see also* Prediction)
   and air pollution, 223, 226 (*see also* Air pollution)

Blood lead level, 225, 226 (*see* Behavior and lead)
Brain, 160, 161

Calcium, 228
Cardiovascular system, 70, 82, 84, 85, 216
Catch-up growth, 100, 101, 104, 116-119, 126, 127, 129, 133-143, 145, 153, 160, 164, 170
Chemicals, 221
   influence of, 211, 221, 228, 237
Chromosomal abnormality, 159
Chronic disease, 99, 133-145
Chronic exposure
   pollutants, 222, 237
   lead (*see* Lead, Blood lead level)
Climate, 172, 235, 240
Coeliac disease, 135-144, 153, 163
Correlation
   in genetics, 49, 51-56, 63-69 (*see also* Heredity)
Co-twin (*see* Twin study)
Cross-sectional surveys, 37-46 (*see also* Survey)
Cultural heredity, 67
Cystic fibrosis, 133-135, 163

Diarrhea, 172
Dominant factors, 63, 66

Endocrines (*see* Hormones)
Endogamy, 86

Environment, 91-94, 100, 170-191, 235-241
  and genetics, 51, 54, 61-69, 73, 75-77, 95, 96, 238
  indices, 56, 57
  influence on growth, 199-206
  medical environment, 99, 100, 145-154, 163-166
Epidemiology, 92, 124, 130, 172, 176, 182
  auxological, 209, 210, 216, 240
  genetic, 69
Epistasis, 63, 67
Error of measurements, 9, 17, 40
Ethnic study, 86-88

Facial measurements, 69, 70
Factor analysis (*see* Heredity)
Fatness, 177
Fitting model, 3, 43
Full-term infants, 109-116 (*see also* Preterm)

Gastrointestinal tract, 216
Generalized distance (*see* Heredity)
Genetics (*see* Heredity)
Gestation length, 102-119, 146-150, 213, 214 (*see also* Appropriate for gestational age, Full-term infants, Preterm, Small for gestational age)

Head
  circumference, 43, 99, 107, 108, 113-119, 127, 129, 147, 148, 150, 151, 153, 160-164 (*see also* Velocity)
  measurements, 74
  shape, 115, 145, 161
Height (*see* Stature)
Heredity, 13, 16, 49-59, 61-77, 91, 92, 95, 109, 113, 151, 152, 162, 163, 174, 182, 184, 238
  component pair analysis, 70
  factor analysis, 69, 70
  generalized distance, 73-74

Heredity (continued)
  linear vectors, 71
  multivariate analysis, 69-75 (*see also* Correlation, Cultural heredity, Environemnt, Epidemiology, Regression, Stature)
Heritability, 51, 53, 56, 63-65, 69, 76
Heterosis, 93
Heterozygosity, 50, 51
Hormones, 5, 32, 83-85, 164, 218
Hydrocephalus, 113

Income level, 179, 182, 201, 203 (*see also* Social influences)
Infectious diseases, 51, 169, 171, 172, 235
Intellectual performance, 151, 153
  and lead, 226 (*see also* Behavior, Mental development)
Intrauterine growth, 100, 101, 123, 145, 147, 151, 159, 172, 212-215

Kwashiorkor, 153

Lactogen (*see* Placenta)
Lead, 224-228 (*see also* Behavior, Blood lead level, Chronic exposure, Intellectual performance, Nutrition)
Life style, 201
Linear vectors (*see* Heredity)
Longitudinal growth, 6, 7, 16, 23-25, 31, 38, 75, 94, 126, 136, 142, 200, 238 (*see also* Mixed longitudinal, Survey)
  missing values, 23-24
Low birthweight (*see* Birthweight)

Malnutrition, 100, 101, 119, 127, 146, 147, 149, 152, 153, 164, 171-177, 235, 236 (*see also* Nutrition)
Matched pairs, 116
Maturation, 24
  skeletal, 170, 235 (*see also* Skeletal age)

Maturation (continued)
  sexual, 43, 45, 92, 93, 100, 161, 164
Maximum a posteriori estimation, 21, 22
Megalocephaly, 113
Mental development, 161, 167, 235 (*see also* Neuropsychology, Intellectual performance)
Mercury, 103
Midchildhood, 7
Midparent value, 3, 4, 7, 31
Mixed longitudinal, 23, 25, 38 (*see also* Longitudinal, Survey)
Morbidity, 42, 93, 94
Mortality
  infant, 149
  neonatal, 42, 145
Multivariate analysis (*see* Heredity)

Neuropsychology, 129, 130, 145, 146, 170, 221, 226, 227, 235 (*see also* Mental development, Intellectual performance)
Neurotransmitter, 216
Noise, 209, 217, 236, 239
Normality, 125
Nutrition, 91, 93, 94, 160–162, 164, 169, 171–175, 184, 186, 189, 191, 200, 203, 235
  high-energy, 118, 119
  maternal nutrition, 54, 94, 175, 176
  susceptibility to lead intoxication, 228 (*see also* Malnutrition)

Oxygen (max. O_2 uptake), 82, 88 (*see also* Respiratory function)

Pair analysis (*see* Heredity, Matched pairs)
Palsy (cerebral), 146
Path coefficient, 56, 67, 68, 69, 76, 179

Peak height velocity, 21, 23, 24, 30, 31 (*see also* Velocity)
Phenylketonuria, 116
Physical activity, 81–88, 166 (*see also* Training)
Physiology, 81–88, 95
Placenta
  modification, high-altitude, 181
  placental lactogen, 214
  placental weight, 148
Plasticity (*see* Adaptation)
Pollution
  industrial, 221–228 (*see also* Air pollution, Stature, Weight)
Polybromated biphenyls
  influence of, 211
Prediction
  Bailey-Pineau, 4
  Bock-Thissen, 3–7, 9–12, 16
  low birthweight, 128, 129
  mature stature, 21, 26, 29–34
  Roche-Wainer-Thissen, 4, 7, 9–12, 22
Preece-Baines model, 21, 24, 25, 32, 75
Prenatal growth (*see* Intrauterine growth)
Preterm, 102, 109, 110, 116, 124 (*see also* Gestation length)
Principal component analysis, 179
Puberty rating (*see* Maturation)
Public health, 37, 42, 93, 97, 123–130, 199, 205, 209, 222

Radiation, 211
Reference curves (*see* Standards)
Regression
  in genetics, 64, 65 (*see also* Correlation, Heredity)
Relationship
  coefficient of, 64–68, 74
Respiratory function, 223 (*see also* Oxygen)

Retardation of growth, 94, 96, 100, 101, 102, 118, 145, 146, 149, 159, 161, 163, 164, 172-174, 177, 179-182

Secular trend, 30, 33, 37, 38, 42, 91-94, 110, 113, 115, 116, 151, 164, 205
Sex development, 99 (*see also* Maturation)
Sex differences, 104, 107, 110, 115
Skeletal age, 3, 4, 9, 12, 13, 17, 31, 181, 224, 225 (*see also* Maturation)
Small for gestational age, 102, 109, 116-119, 123-130, 145-149, 160, 162, 165 (*see also* Appropriate for gestational age, Gestation length, Full-term infants, Perterm)
Smoking effect, 169, 180, 201, 202, 235
Social influences, 33, 40, 41, 52, 56, 57, 69, 91-94, 97, 125, 146-151, 159, 161, 165, 169, 175, 179-180, 182-184, 189, 190, 199-208, 212, 213, 223, 226 (*see also* Income level, Stature)
Standards, 37, 38, 40, 46, 94, 95, 97, 124, 159, 163, 164
Stature, 31, 83, 93, 99, 104-107, 110, 112, 113, 117, 127, 133-145, 148, 151-153, 160, 161, 164, 236
  genetics, 49-59
  and pollution, 224, 227
  and social factors, 38, 42, 44, 45, 170, 179, 183-190, 200-206 (*see also* Velocity)
Step-test, 87
Stress, 169, 170
Supine length (*see* Stature)

Survey, 37-46, 91, 95 (*see also* Cross-sectional surveys, Longitudinal, Mixed longitudinal)

Teeth, 166
Thoracic diameter, 127
Training, 82-86 (*see also* Physical activity)
Triple-logistic, 4-7, 12, 17, 18, 21, 22, 25, 29-34
Twin study, 65-66, 70-75, 92, 95, 163
  co-twin, 82-86

Ultrasound, 101, 106, 160
Urbanization, 185, 187-189, 204, 205, 211, 212, 236, 238, 239

Velocity, 100, 160, 161, 164
  head circumference, 107, 108, 116
  supine length, 6, 106, 107, 113, 136-143
  weight
    full-term, 110, 125-126, 136-143
    intrauterine, 104, 105
    (*see also* Peak height velocity)
Vitamine D, 49, 223

Weight, 42, 44, 45, 93, 103-106, 110, 111, 117, 133-144, 148, 151-153, 159, 161, 164, 170, 179, 181, 185, 186, 235
  fetal weight, 104
  for height, 42, 44, 45, 161, 162
  pollution 224-227 (*see also* Birthweight, Placenta, Velocity)

Zinc, 177

DATE DUE